In Search of a Theory of Everything

In Search of a Theory of Everything

The Philosophy Behind Physics

DEMETRIS NICOLAIDES

OXFORD
UNIVERSITY PRESS

Oxford University Press is a department of the University of Oxford. It furthers the University's objective of excellence in research, scholarship, and education by publishing worldwide. Oxford is a registered trade mark of Oxford University Press in the UK and certain other countries.

Published in the United States of America by Oxford University Press
198 Madison Avenue, New York, NY 10016, United States of America.

CIP data is on file at the Library of Congress
ISBN 978–0–19–009835–3

3 5 7 9 8 6 4 2

Printed by Integrated Books International, United States of America

For my Maria-Christina, my daughter, the most precious of all that exists!

For Anna, my loving wife

For my mother and the memory of my father, for their unconditional love

Contents

Acknowledgments   ix

    Introduction   xi

1. Philosophy and Physics   1

2. Close Encounter of the Philosophical Kind   5

3. The Quest for a Theory of Everything   7

4. Cosmic Justice   18

5. The Stepping-Stone to Truth   30

6. Numbers and Shapes   36

7. The Changing Universe   54

8. The Unchanging Universe   72

9. Paradoxes of Nature   87

10. The Chemistry of Love and Strife   99

11. In Everything Is Everything   110

12. Atoms of Matter and Energy   120

13. Atoms of Space and Time   135

14. It's Fate—Maybe   152

15. Atomic Connections   157

    Epilogue   162

Bibliography   163
Index   167

Acknowledgments

I thank my teachers—including my PhD mentor, professor Alexander A. Lisyansky, and professor Jacob Neuberger—for teaching and challenging me, and my students, whose thirst for knowledge keeps me improving.

I'm grateful to physicist Walter Polkosnik, philosopher Richard D. McKirahan, and historian of science Joseph D. Martin, for their useful comments and overall support of this book project.

I'm sending special thanks to an anonymous Oxford University Press reviewer, whose insightful reading of the manuscript and subsequent thoughtful comments and recommendations undoubtedly led to a significant improvement of the book.

I am forever grateful to my literary agent, Nancy Rosenfeld, for her kindness, wisdom, guidance, and for giving hope to the writers of the world.

I'm thankful to Bloomfield College for giving me the opportunity to research this topic.

All people from Oxford University Press deserve a special recognition for their diligent work. Those with whom I had the pleasure of interacting directly include editor Jeremy Lewis, editorial assistant Bronwyn Geyer, project manager Rajesh Kathamuthu, and editor Leslie Johnson. My sincere thanks!

I'm thankful to my wife, Anna, for her continuous support in pursuing my research, and to our daughter, Maria-Christina, for asking so many questions about the world—some of which have found their way into this book.

Introduction

In Search of a Theory of Everything is an adventurous journey in space and time in search of a unified "theory of everything" by means of a rare and agile interplay between the natural philosophies of influential ancient Greek thinkers and the laws of modern physics. For a theory of everything, all the phenomena of nature share a subtle underlying commonality and are explainable by a single overarching immutable principle. Reading the past for what it is, is of tremendous value, but so is its reading from the perspective of modern knowledge. Not to judge it for its flaws but to be inspired by its insights. This comparative study of the universe is the spirit of *In Search of a Theory of Everything*—to physics through philosophy, to the new via the old, in a balanced way.

A relatively "easier" analysis of nature, that of a major natural philosopher of antiquity, commences every chapter to fasten the bedrock for the more complex. The transition into the more complicated views of modern physics is gradual and systematic, entwining finely the two, the ancient with the new, the forgotten with the current, by unfolding a history and a philosophy of science, and connecting all the great feats of the mind and time. Those philosophers had ideas that resonate with aspects of modern science; puzzles about nature that still baffle;[1] and clever philosophical rationales that can be used to reassess completely anew fundamental but competing principles of modern physics, even to speculate about open physics problems. *In Search of a Theory of Everything* is a new kind of sight, a philosophical insight of modern physics that has long been left unexamined.

About 2,600 years ago, the ancient Greeks had a magnificent intellectual awakening that contributed to the rise of their civilization. "Suddenly" the age-old popular mythological worldviews were questioned, rethought, and eventually changed. Nature was no longer seen as a chaos of random, unpredictable, and incomprehensible phenomena attributed to mysterious supernatural forces through myths, superstition, and the chancy decisions of capricious, anthropomorphic gods. On the contrary, nature was viewed as a cosmos: a well-structured, organized, ordered, harmonious, self-contained, self-consistent, and beautiful whole in which the phenomena were natural components that obeyed intrinsic causal laws that could be discovered and understood by the practice of rational analysis of nature and without invoking the supernatural. A profound transition

---

[1] Carlo Rovelli holds the same view: Carlo Rovelli, *Reality Is Not What It Seems* (New York: RiverHead Books, 2017), 21 (Kindle ed.).

in human thought took place that was a consequence of the realization by these philosophers that nature *is* comprehensible. A simple question emerged: what is the nature of nature? The prolific answers these Greek thinkers offered ascribed purely naturalistic causes to all phenomena in nature and created a new outlook, the scientific! Science has been influencing the world ever since by guiding it out of the cave of ignorance and into the light of truth.

1
Philosophy and Physics

Episteme means "knowledge" in Greek. For Aristotle, we have episteme if we know the cause of something.[1] *Science* is episteme in Latin. Hence, strictly speaking, science includes all fields of knowledge (theology, philosophy, physics, history, etc.). Physics is a particular type of science: it is the study of *physis* (nature in Greek). With time, however, especially nowadays, the word *science* evolved to have a narrower focus, and it, too, means the study of nature. Because of this, physics and science have basically become synonymous, but also so because all subfields of science that study nature are really reducible to the natural laws of physics. For example, biology studies cells, which are made of molecules; chemistry studies molecules, which are made of atoms; and physics studies atoms. Therefore, generally all the sciences about *physis* are ultimately built from (and are branches of) physics.[2] The natural philosophers[3] of antiquity that we'll consider were philosophers but also physicists. Philosophy and physics were then closely related fields.

According to legend, Pythagoras probably coined the term *philosophy*, "love of wisdom," in Greek. So to define *philosophy* we need to know what wisdom is. Recall that the Oracle of Delphi prophesied that Socrates was the wisest, but he humbly doubted it.[4] Using the art of dialectic (his famous questioning style, the *maieutic*), he set out to prove the Oracle wrong (by engaging in meaningful discussions with his fellow Athenians, politicians, poets, craftsmen, and farmers, in search of the truth) only to find out that the prophecy was spot-on. He was the wisest indeed for, unlike the others, who confused their skill with wisdom, their *empeiria* (in Greek, experience or talent) with episteme,[5] he at least knew well one thing: that he knew nothing.[6] Confessing ignorance gives healthy skepticism hope to advance to the truth and become wise. And the road toward wisdom begins for Socrates with the "know thyself,"[7] by recognizing our mind's limitations and the uncertainties of our methods of inquiry.

---

[1] Aristotle, *Organon*.
[2] The word *science* will have this narrower meaning from this point on.
[3] Called so because they investigated nature.
[4] Plato, *Apology*, 20e–21b.
[5] They also confused *gnome* (judgment or opinion, not the garden ornament) with *gnosis* (knowledge).
[6] The exact quote is: "I know one thing; that I know nothing."
[7] Inscription found at the Oracle of Delphi in Greece.

I want to say, parenthetically, that knowing *nothing* is inconceivable. For the notion of "nothingness" is not allowed; it's an impossibility! In chapter 8, we'll talk about nothing, in reference to the philosophy of Parmenides—the philosopher who contemplated well, nothing, first—in order to check the validity of one of Einstein's most bizarre claims, which he based on the physics of his relativity, that, change, in the universe, is an illusion.

Anyhow, I'm not sure what wisdom might be, but I'm sure that the journey to it begins with a question and continues with courage, determination, and an honest effort to eliminate age-old, false beliefs and prejudices and seek rational (not dogmatic) answers. Now, although we are born imaginative creatures with insatiable curiosity and desire for adventure and knowledge, with age we usually settle down both physically and mentally (our body and mind) and inquire less and less. Asking is a trait of youth but of philosophy, too. Neither solar nor entropic[8] time ever ages the inquiring mind.

In a more concrete example, philosophy for me is this: I read a book about physics (of Albert Einstein, Steven Hawking, Michio Kaku), and I usually understand it well. Then I read a book about philosophy (of Aristotle, Bertrand Russell, Karl Popper), and I often don't understand it that well. So then I reread it, and again, only to find out that I now understand better the *physics* book. And so sadly I keep on reaffirming that "the man of science is a poor philosopher."[9]

Philosophia[10] is She who managed to escape from the darkness of superstition and ignorance (from Plato's cave) and dared to ascend into the world of light and knowledge. It is "the vision of truth,"[11] it is the ability to "teach people to talk sense,"[12] it is sameness in dissimilarity, or a subtle immutability in conspicuous change. It is part Apollo (reason) and part Dionysus (passion) but never any one alone, it "is something intermediate between theology and science,"[13] or, put simply, philosophy might be what the Greeks called "the gift of wonder"[14]—of imagining, searching, discovering, and learning all you can by wondering.

Adding to such gift is science, the systematic study of nature and the organization of acquired knowledge into "timeless, universal," causal, and, most

---

[8] The law of entropy describes the universe's tendency toward greater disorder. It is manifested in everything, in humans, too. With time our physical attributes grow more disorderly, and we basically age. Eating slows down the increase in entropy and thus the aging process. I'm implying that thinking, too, may slow down the aging process.

[9] "400 Albert Einstein Quotes That Will Move (and Surprise You)," Wisdom Quotes, http://wisdomquotes.com/albert-einstein-quotes/ (accessed July 14, 2019).

[10] Philosophy in Greek.

[11] Plato, *The Republic*, Book V, trans. Benjamin Jowett, in *The Complete Works of Plato* (The Complete Works Collection, 2011), Kindle Locations 25382–25383.

[12] Karl R. Popper, *Conjectures and Refutations: The Growth of Scientific Knowledge* (London: Routledge, 1989), 68.

[13] Bertrand Russell, *The History of Western Philosophy* (New York: Simon & Schuster, 1945), xiii.

[14] Popper, *Conjectures and Refutations*, 72.

important, testable laws that are derived from observation and rational consideration. A good scientific theory, therefore, makes experimentally verifiable *and* falsifiable predictions, which must be tested by experiment. Science is evidence-based knowledge; it is not knowledge based on opinion or dogma. The defining characteristic of science is its unique way of study, the scientific method. It can be summarized in five steps.

(1) Observe nature. For example, things fall. (2) Formulate a question, a problem (based on observation). Why do things fall? (3) Answer the question with a hypothesis (an educated guess) that makes *testable* predictions (the cardinal rule of science). They fall because the earth is pulling them: dropped from rest, they should fall 4.9 meters in 1 second. (4) Perform properly designed reproducible experiments to collect data that will be used in order to verify or falsify the predictions of the hypothesis. Drop various objects from rest and measure the distance they fall in 1 second. (5) Draw conclusions by comparing the predictions of the hypothesis against the data of the experiment: (a) if the predictions are observed (i.e., they agree with the data, so things would indeed fall 4.9 meters in 1 second), the hypothesis is verified and it transitions into a scientific fact, a law of nature; (b) if the predictions are not observed (they are in disagreement with the data, things would not fall 4.9 meters in 1 second), the original hypothesis is falsified and thus replaced by a modified one or by something completely new.

This is the general scientific method as was formulated by philosopher Francis Bacon. A clarification: as Einstein, Popper, Richard Feynman, and Hawking argued, we can only verify a hypothesis; we can't prove it—we can never be *absolutely* certain if the laws we discover are truly timeless and universal. For example, while experiment after experiment keep on verifying Einstein's relativity, "No number of experiments can prove me right; a single experiment can prove me wrong," said he.[15] That is, a new type of experiment may find a flaw in the theory (not previously detected) that will prompt scientists to revise or completely change it with a new vision.

In the quest for truth about nature, science without (the wisdom of) philosophy is practical and rational but (arguably) dull, and philosophy without (the empirical facts of) science is abstract and wise but (experimentally) unverified.[16] For Einstein, "science without epistemology is—in so far as it is thinkable at all—primitive and muddled."[17] The road to truth, I believe, is paved by science and

---

[15] Quoted in Paul G. Hewitt, Suzanne Lyons, John A. Suchocki, and Jennifer Yeh, *Conceptual Integrated Science*, 2nd ed. (Boston: Pearson, 2013), 5.

[16] At least in the sense that philosophers in general don't test their own claims via experimentation; rather, they rely on the soundness of their logic. But for physicists it is experiment that ultimately defines what constitutes scientific knowledge.

[17] Abraham Pais, *Subtle Is the Lord: The Science and the Life of Albert Einstein* (New York: Oxford University Press, 2005), 13.

philosophy, but certainly by other fields of knowledge, too. The view, otherwise, is of mere copies, shadows of truth.

In Plato's parable of the cave,[18] prisoners constrained in a cave since childhood and for many years thereafter can see only shadows and mistake these shadows for the only reality. The prisoners therefore are utterly ignorant of the objects that project the shadows. But the sight and insight of the prisoner who eventually manages to escape start to gradually improve. At last, outside the cave and in the light of the sun, she begins to have a better perception of reality. She now sees how things resemble their shadows, and she realizes that shadows are deceptive; they are only a mere copy of the real objects (and of the grander truth in general). Nature, she now knows, is much more than what She appears, but Her divine secrets can ultimately be untied by the curious, willing mind. Eager to share her newfound sense with her prisoner-friends, she descends once more into the cave. But passing suddenly into the darkness from the light makes her sight briefly feeble. Her friends are tricked by it and think, up she went with her eyes and down she came blind; thus, sadly, they believe that the cave is the only safe place.

---

[18] Plato, *Republic*, Book VII.

2

Close Encounter of the Philosophical Kind

It is the beautiful season of summer. I have been reading this great story for hours, since dusk and during the absolutely moonless night. It is now almost dawn. I am very tired but do not want to put the book down. It is unusually original and wonderfully profound. The pages are one by one turning. I am really exhausted and am fighting sleep because the story is so good.

All of a sudden I hear a splash. The day is hot and bright, the sky blue, the sun yellowish-white. I look up and see old sage Thales playing, falling into the fresh, cool, flowing waters of a river, in search of something. "Not very wise," the atomic Democritus remarks with a smile, staring curiously in the void with undivided attention. He is vastly confident. Quiet Leucippus, his teacher, is solidly standing by. But also because moments later, pleasant Epicurus, the student of them both, hopping from his garden he swerved by the scene by his own free will. Brilliant Aristotle is meticulously studying them all, but he is not intimidated at all. *His* teacher is *Plato*, the philosopher of Ideas or Forms. Where is he anyway? He is usually all over the place, but don't ask Socrates, his teacher; he knows nothing.

In the meantime, Thales is carefree and zestful, having the matter of water primarily under control, getting up but purposely jumping back in the river again, with novel, childlike, passionate, and playful curiosity. Off from the center of the action, revolving around a Central Fire, carefully preparing his food while singing in a harmonious but almost secretive whisper, is the legendary Pythagoras. Inspired by the moment, he stops the song and begins counting the proportionally spaced ripples of the water from each splash Thales makes. He is a prominent mathematician, able to face squarely all mathematical irrationalities, but he is also a cosmopolitan musician who enjoys master-like attention from orderly and exclusive gatherings of crowds. "Cosmic justice, conserve and save the phenomena," thirstily shouts the infinitely abstract but also practical Anaximander, the genius of antitheses, feeling the heat of dry air and expecting that it will soon be neutralized by the opposite coolness of wet water.

It is really a beautiful and hot day, but imaginative Anaximenes, leaping over a stepping-stone, finds a creative and concrete way to moderate the heat and thereby cool down. With his lips nearly closed, he blows air out onto his body, noting that it emerges colder than when his mouth was wide open, causing his condensed sweat to rarefy and evaporate. Sitting at a distance, away from the many, boldly being where no one has been before, is the enigmatic Heraclitus, who skeptically observes the process of the constantly changing events, going through conspicuous but also subtle changes. He is quite certain he has previously taken a bath in this river's fresh waters, but then again, strangely, everything looks new and changed. What is the Logos (cause) of all these eventful processes? To the contrary, judging all of the sense-perceived reality to be deceptive, there is the one and only Parmenides the ontologist, proud and relieved. For journeying during the darkness of night and into the light of day, through the unknown, from afar, he found the true way here by intentionally avoiding the known and opinionated way of all others.

Anaxagoras's *nous* (or intellect) finds everything, in *everything*, to be a puzzle: "How is it that all these people from different eras of time and different places are here?" He wonders by skillfully placing his hands over his head. "Indeed a paradox, a paradox of space and time," adds the argumentative and prolific Zeno, who, through dialectic (the method of reductio ad absurdum, or reducing to the absurd), is trying to prove that motion is an illusion of the senses, and so no one really moves, despite that all appears to so do. "Are you sure space and time are the only elements in the puzzle?" melancholy Empedocles challenges, while, in the name of episteme and his love for strife, holding tight onto his clepsydra (a water clock), he risks a dangerous experimental leap through the Air and over the flames of Fire but lands safely on Earth, in fact in the Water, just beside me.

"And who might you be, young fellow—the modern physicist?" he asks. As I respectfully nod in awe, I feel all eyes curiously staring at me as if I'd been expected. And immediately the brightness of the day surprisingly turns into a mysterious twilight. "I have been predicting an eclipse at your arrival," Thales says, while nostalgically shaking off primarily the substance of water from his wet, muddy, ripped, and unfashionable clothes. "We have been longing to know your story," he adds. Moments later, the bright daylight is pleasantly restored. It is now noon. "It is a beautiful day indeed," I say humbly, "for I am learning yours. And I will tell you mine, too, but under *your* sunlight, for eclipses are ephemeral and pass, but your knowledge is timeless. You *still* bring fire to modern science." The day is still young, and who knows of the morrow?

Everyone's senses are keen, observing the changing sights, listening to curious sounds, smelling soul-awakening aromas, tasting the sweet air, touching the cosmic elements. But so is everyone's intellect contemplating it all. What a beautiful day! What a beautiful nature! What is her nature?[1]

---

[1] This chapter was inspired by chapter 2 of *The God Particle*, in which Leon Lederman imagines conversing with Democritus: Leon Lederman and Dick Teresi, *The God Particle: If the Universe Is the Answer, What Is the Question?* (Boston: Houghton Mifflin, 1993).

3

The Quest for a Theory of Everything

Introduction

Thales (ca. 624–ca. 545 BCE) was interested in how nature works. He was the first to ask what things are made of and what the properties of matter are. These are still the most fundamental and difficult questions of science. His answers were based on solely rational arguments, uncluttered by myths, superstition, rituals, or the actions of capricious gods. His approach was therefore the same as that of modern science.

He reasoned that in spite of the apparent diversity and complexity in nature, all things are made from the *same* stuff (water), and all things obey a common set of unchanging basic principles (water's transformations, e.g., its solidification, liquefaction, and evaporation). Thus, for Thales, nature is characterized by a certain sameness or unity between all things, however diverse they may be, an overall intrinsic simplicity.[1] Thales's quest for sameness is modern physics' search for a unified theory of everything. It tries to unify the four fundamental forces of nature—the electromagnetic, the nuclear strong, the nuclear weak, and gravity—and discover the one primary substance of matter from which everything is derived. A major challenge of this undertaking is to find a quantum version of gravity, a most difficult task.

What Are Things Made of?

We still don't know what things are made of. Nonetheless, presently, according to the standard model of physics (introduced later in this chapter), things are not made from water but from microscopic particles called quarks (constituents of protons and neutrons) and leptons (particles including electrons). And the plethora of diverse things is partly due to their transformations (from one type of particle into another), not to the transformations of water.

---

[1] Aristotle, *Metaphysics* 983b6–13, 17–27. Or see Daniel W. Graham, *The Texts of Early Greek Philosophy: The Complete Fragments and Selected Testimonies of the Major Presocratics* (Cambridge: Cambridge University Press, 2010), 29 (text 15); Aëtius 1.31, 1.10.12. Or see Graham, *Texts of Early Greek Philosophy*, 29 (text 16); Simplicius, *Physics* 23.21–29. Or see Graham, *Texts of Early Greek Philosophy*, 29 (text 17).

Hence, both Thales and the modern physicist are wrong but right, too. They are wrong because neither water nor quarks and leptons are the primary substance of matter—what is, we are still in search of. But they are also right because all things in nature, evidence suggests, share a subtle underlying common law and are made from the transformations of one and the same substance, regardless of how different everything seems. His idea about the transformations of matter, in particular, not only describes a fundamental property of the modern concept of energy (or matter, since, as Einstein's special relativity theory makes clear, they are equivalent and transmutable into each other), namely its ability to transform into various forms and cause change, but also employs causality (a relation between cause and effect), because for him the cause of all other things is the transformation of just one primary substance. But why was water the underlying principle/cause of such sameness and unity?

Why Water?

Several observations might have stimulated Thales in his speculation that all things are transient forms of water. Some ancient accounts such as Aristotle's[2] and Aëtius's[3] give us some insight. Water is required for the survival and development of all kinds of life. Primitive life exists in moist environments, and animal sperm is liquid. Also, since water transforms easily into the three forms of matter, the solid (as ice), the liquid, and the gaseous (as water vapor), and into a variety of shapes, it could, Thales might have thought, also transform into everything else, such as rocks or metals. Now, while all substances transform into the three states of matter (e.g., given enough heat, a solid piece of metal can melt and evaporate), water is the only substance of daily experience that does this before our eyes and on a regular basis through the changing seasons, something that observant Thales could not have missed. Furthermore, it transforms more easily: its evaporation temperature of about 100 degrees Celsius (at sea level) is smaller than that of, say, copper, bronze, or iron—materials that in antiquity were heated and melted to make tools—so one does not need a lot of heat to vaporize it; and in the cold winter, water is the only substance to turn to snowflakes and solid ice of all sorts of shapes. So its choice as a primary substance over other things appears logical. Thales might have reinforced his water hypothesis, I speculate, from another everyday observation, namely, that when heated or burned, all things release (or so it seems, anyway) water vapor. One example that might have been an

---

[2] Aristotle, *Metaphysics* 983b6–13, 17–27. Or see Graham, *Texts of Early Greek Philosophy*, 29 (text 15).
[3] Aëtius 1.31, 1.10.12. Or see Graham, *Texts of Early Greek Philosophy*, 29 (text 16).

inspiring clue for his water doctrine might be observing the rising smoke from a burning piece of wood mixing with air and clouds, which in turn can be mixed with rain or snow and blend with the soil on earth. At first glance, smoke, air, and clouds, are like (or seem to be) water vapor; rain and snow are water, and soil contains the water and snow of the rain or snowstorm. Therefore, soil might be thought of as transformed water, and so might then be the plants (thus wood), since, starting as seeds, plants grow from the soil and "are nourished and bear fruit from moisture,"[4] and so might also be the animals, since they eat plants or each other.

Since processes of this sort appear causal with water as the first cause, then it seemed logical to assume that everything is made from the same stuff— reconstructed from the same first principle—and that, in general, everything in nature is characterized by a certain subtle sameness.

The Quest for Sameness

Sameness is a core concept in modern physics, not only because it emphasizes a universal, underlying, simple principle as a characteristic of all things in nature, but also because it points to a commonality in their ultimate origin. Unity (in the sense that everything can be derived from one and the same principle), Thales reasoned, is a subtle, intrinsic property of nature.[5] This idea inspired all natural philosophers (each creating his own special theory on unity, as we'll see in subsequent chapters), and, in turn, they have inspired scientists of recent times in their own searches for a theory of everything.

A Brief History of Recent Times

James Clerk Maxwell (1831–1879) unified successfully the electric and magnetic forces by proving mathematically that they are really two manifestations of the same force, the electromagnetic. The electric force is caused by the electric charge: the positive and negative. Objects of opposite electric charge attract one another while objects of the same type of charge repel. The magnetic force is caused by an electric charge in motion. A permanent magnet has two poles, the north and the south. Opposite magnetic poles attract, and similar magnetic poles repel. He also unified electromagnetism with light. An electron oscillating

---

[4] Ibid.
[5] Ibid.; Aristotle, *Metaphysics* 983b6–13, 17–27. Or see Graham, *Texts of Early Greek Philosophy*, 29 (text 15).

up and down produces an electromagnetic wave, *light*, just as a floating cork oscillating in and out of water produces a water wave.

Albert Einstein (1879–1955) from 1925 until 1955 attempted unsuccessfully to unify the electromagnetic force with gravity. Gravity is still the most puzzling of the forces, although it was the first force to be described mathematically, with the law of universal gravitation of Isaac Newton (1642–1726), and advanced significantly through Einstein's theory of general relativity.

Nonetheless, success struck another physics front with the combined efforts of Sheldon Glashow (1932–), Steven Weinberg (1933–), and Abdus Salam (1926–1996). In the 1960s, the three physicists managed the unification of the electromagnetic force with the nuclear weak force in what is known as the electroweak force.[6] The nuclear weak force is responsible for the radioactive decay of unstable nuclei such as that of uranium, and the transformation from one type of material particle into another—Thales's notion on the transformation of matter is a significant process in modern science. The experimentally confirmed unification of the electromagnetic force with the weak force occurs at high energies and temperatures—where the two forces have the same strength and are indistinguishable; thus, they are considered as one force. However, at lower energies/temperatures (generally those of everyday experiences), these two forces are two expressions of the same force: the electroweak.

The standard model of physics is the theory that combines the knowledge of the electroweak force and the nuclear strong force—which binds the quarks in the protons and neutrons and also the protons and neutrons in the nucleus of an atom. It is the best model so far because it combines successfully several theories to explain how particles interact and how the universe works. Quarks and leptons are among several experimentally confirmed predictions of the standard model. In fact, even more important with respect to Thales's view, according to the standard model, the materialness of quarks and leptons—in particular the source of their mass—is *one and the same* type of particle, the famous Higgs boson. It was discovered in 2012 at the Large Hadron Collider, the most powerful atom smasher in the world. Although successful, the standard model has a few major challenges: it doesn't include gravity and can't explain dark matter, dark energy, and why there is more matter than antimatter in the observable universe—topics to be discussed later.

Through a grand unified theory (GUT) physicists hope to extend the standard model by creating an experimentally verifiable theory in which the electroweak force and the nuclear strong force are unified. Several good candidates for a GUT do exist, making concrete testable predictions (such as the decay of a proton, not

---

[6] Stephen Hawking, *A Brief History of Time: From the Big Bang to Black Holes* (New York: Bantam Books, 1988), chap. 5.

yet observed), though none has so far been experimentally verified. It is hypothesized that these forces were indistinguishable only for a miniscule moment, 10^{-35} seconds after the big bang, when the universe was superhot. According to the big bang cosmological model, about 13.8 billion years ago the entire universe was unimaginably small, possibly a mere point, infinitely dense and hot. It then exploded in the absolutely most extraordinary event called the big bang and has ever since been expanding, cooling, rarefying, and creating the eventful universe we live in. The idea of the big bang originated with Georges Lemaître (1894–1966), a Belgian priest trained in physics who used Einstein's relativity to predict it.

Finally, a community of ambitious physicists is currently on the quest for the ultimate principle of sameness, that is, for the absolute unification of all four aforementioned fundamental forces of nature in what is termed the theory of everything (TOE). A TOE hopes to establish that everything in nature is explainable by a single overarching timeless principle and its associated equation, that everything is truly a consequence of just one primary substance, as Thales initially envisioned, and one force—why four forces? Oneness, it appears, has been evolving to be a simpler, more preferred philosophy, for both our science and our religion (i.e., monotheism).

The Challenge for a Theory of Everything

Through a TOE we hope to reduce all of nature into one fundamental force and one fundamental substance with its transformations, an utter simplicity and sameness of Thalesian grandness. The paramount challenge in finding a TOE is rooted in our inability so far to find a quantum version of gravity, commonly called "quantum gravity." That is, to combine the rules of quantum theory (also known as quantum mechanics or physics, and which includes the standard model) with the rules of Einstein's theory of general relativity—or find new rules *completely*. Since nature is one and beautiful, one and beautiful should also be the theory that would explain Her. But the two theories we have for Her are mutually exclusive, a very displeasing situation in science. Probability reigns in quantum mechanics, determinism in relativity. Nature is granular in the first theory, smooth in the second. Space is a "mute" immutable container in quantum (and Newtonian) physics for things to just be in, but a dynamic malleable fabric in relativity that "tells matter how to move."[7] A cosmic clock tells the same exact time for everyone everywhere, objects have a fixed size, and time travel is impossible

---

[7] Charles W. Misner, Kip S. Thorne, and John Archibald Wheeler, *Gravitation* (Princeton, NJ: Princeton University Press, 2017), 5.

in the former theory; but in the latter theory, time is relative (it slows down or speeds up depending on how you move or where you are), moving objects contract, and you can travel into the future. Quantum theory describes successfully the world of the tiny, of atoms, electrons, protons, neutrons, quarks, and so on. Relativity describes successfully the world of the large, of planets, stars, galaxies, and generally the large-scale universe by explaining how space, time, matter, and energy are all inextricably intertwined and how gravity works. (Aspects of both of these theories will be discussed in later chapters.) Quantum theory and general relativity are significant improvements over Newton's and Maxwell's physics. Nevertheless, as special cases of the former two, the latter two are still abundantly practical.

String Theory

A possible TOE is string theory. It is highly speculative and without yet any experimental support for its claims. It seeks to describe nature in terms of vibrating strings of energy in eleven dimensions. According to string theory and in agreement with the essence of Thales's idea, everything is made from the same stuff: absolutely identical strings, theorized to be the primary substance of the universe. These strings, like violin strings, have different modes of vibrations that are speculated to manifest as different types of particles, which include the quarks and leptons.[8] Now, we are aware of the three dimensions of space (think of them as three edges of a cube meeting at a vertex) and the one of time—thus with respect to what we can readily experience, we live in a four-dimensional universe. The other spatial dimensions predicted by string theory are hypothesized to be curled up into unimaginably small, ball-like geometrical shapes of size 10^{-35} meters, known as Planck length, as big as the vibrating strings themselves, and thus not easily detectable.

Loop Quantum Gravity

Like string theory, loop quantum gravity is a theory of quantum gravity—it attempts to combine the rules of quantum physics with those of relativity. Although loop quantum gravity is not a TOE (for it doesn't try to unify the four

---

[8] Brian Greene, *The Elegant Universe: Superstrings, Hidden Dimensions, and the Quest for the Ultimate Theory* (New York: W. W. Norton & Company, 1999), 144; Michio Kaku and Jennifer Trainer Thompson, *Beyond Einstein: The Cosmic Quest for the Theory of the Universe* (New York: Anchor Books, 1995).

forces as string theory does), its findings might still help to find one. But like string theory, loop quantum gravity too is very speculative without experimental backing. Loop quantum gravity seeks to describe nature by radically reimagining space.[9] Space is not composed of points—it's *not* infinitely divisible into ever-smaller regions. Space is composed of *magnitudes*: indivisible, interlinked, finite space expanses—"atoms" of space—shaping like loops, or rings, roughly of Planck length. The word *atom* means "indivisible" in Greek, implying that there is a smallest "cut" of something; thus, that cut cannot be divided further. The notion of an atom is of central importance to all of science. Atoms of matter have first been hypothesized by Leucippus and his student Democritus (chapter 12). And atoms of space and of time have been the innovation of Epicurus (chapter 13), a student of their atomic school of thought. Atoms of matter have of course been experimentally verified, but not the atoms of space and time.

Black Holes: Challenges in the Quest for Sameness

A black hole is a supermassive, dense, point-like object with immense gravity. The event horizon is the invisible spherical boundary around it, from within which nothing, not even light, can escape. Thus, the region in space enclosed by the event horizon is black. Since light can't escape, events that might be occurring within a black hole can't be seen by us. Consequently, if some kind of civilization existed there, its citizens could see us (as light from us enters the back hole), but we couldn't see them. Black holes exist in every galactic center, although when they were first predicted by Einstein's general relativity, Einstein himself dismissed their existence as mere mathematical artifacts. The first-ever image of a black hole was released in the spring of 2019.

Now, a black hole is a *tiny* object with *enormous mass*—thus both quantum mechanics and relativity apply. But the properties of black holes are different for each theory.

For example, according to relativity, a book falling into a black hole is crossing a calm event horizon as if it were nothing special to pass through. Eventually, the book is ripped into pieces by the immense gravity and is crushed at the infinitely dense center of the black hole. What once was a distinguishable book is now just matter indistinguishable from all other matter already there. And all information about it is thereafter lost forever.

But the falling book, according to quantum mechanics, is crossing a highly energetic event horizon, a region of fire that burns the book as it goes through. Still

---

[9] Carlo Rovelli, *Reality Is Not What It Seems* (New York: RiverHead Books, 2017) (Kindle ed.); Carlo Rovelli, *Seven Brief Lessons on Physics* (New York: RiverHead Books, 2016) (Kindle ed.).

though, some energy is radiated[10] to us from the region just outside the black hole that amazingly contains subtle information about the fate of the book and all its contents while inside the black hole. Information therefore is preserved.

These conclusions clearly clash: if information is lost, quantum theory is fundamentally wrong, but if information is preserved, it is general relativity that is fundamentally flawed. This contradiction is known as the information paradox, and it is still an open question. General relativity and quantum theory don't see eye to eye in explaining black holes, but a TOE should lift the contradiction.

The Sage

Thales was regarded as one of the seven sages of the ancient Greek world and the wisest among them. For this he was offered a golden cup, which he respectfully declined by offering it to another of the sages, who offered it to another until the cup was again returned to Thales. He then dedicated it to the god Apollo at Delphi. Thales, a humble man, is credited (among several other Greeks) with the famous aphorism "Know thyself." He was a philosopher, scientist, astronomer, mathematician, politician, and even a theologian known best for his belief in hylozoism, that "all things are full of gods."[11]

But he was also a practical man. As an engineer, for example, he aided the army of King Croesus of Lydia in crossing the Halys River by digging a deep trench in the shape of a crescent and diverting its waters. It was a Herculean feat indeed. The waters initially flowed by one side of the army but later diverted; they flowed by the opposite side. Through his observations of the night sky, he discovered the stars of the Little Dipper (Ursa Minor), which includes the North Star, Polaris, and used them to teach navigation. He also wrote treatises on various calendars such as on the spring and fall equinoxes, on the summer and winter solstices, on the phases of the moon, on solar eclipses, and on the rising and setting of certain stars such as the Pleiades.

While in Egypt Thales is said to have computed the height of a pyramid by first noticing that at a certain time of day his own shadow was as long as his height. He then concluded that the length of the pyramid's shadow at that same time of day was, according to the law of similar triangles, equal to the pyramid's actual height. While on land and through the use of geometry, he was able to calculate his distance from a ship at sea. Furthermore, he estimated correctly the angular size of the sun and of the moon relatively to the angular size of their apparent orbit in the sky to be equal to 1/720.[12] Angular size of an object is the

---

[10] Hawking, *A Brief History of Time*, chap. 7 ("Black Holes Ain't So Black").
[11] Aristotle, *On the Soul* 411a7–8, trans. Graham, *Texts of Early Greek Philosophy*, 35 (text 35).
[12] Diogenes Laërtius 1.24. Or see Graham, *Texts of Early Greek Philosophy*, 21 (text 1).

angle created from your eye to two diametrically opposite points on the object. Say the dot, •, represents the eye, letter I, the object, and that they are situated like so, • I, from each other. The angular size of I is the angle depicted in the following geometry: •<I. Today we know that the angular size of both the sun and of the moon is 0.5 degrees (a fact that is true because the sun is much farther away than the moon). Incidentally, had these angular sizes happened to be unequal, their apparent sizes would also have been unequal, and, consequently, a total solar eclipse, during which our view of the sun is completely covered by the presence of a new moon in between (like the one Thales predicted on May 28, 585 BCE that stopped that day's battle between the Lydians and the Medes, and their six-year war),[13] would not have been possible. Now the sun's and moon's angular size of 0.5 degrees divided by 360 degrees, which is the angular size of their apparent orbit around earth, is exactly equal to Thales's estimation, namely, $0.5/360 = 1/720$!

Thales was also known for his weather predictions, a skill proven valuable in teaching his fellow citizens an important lesson about life regarding their negative attitude toward philosophy. In spite of all his knowledge (practical and abstract) and all his wisdom, Thales is said to have been poor. And because of his poverty, some people criticized philosophy by calling it a useless and impractical way of life. According to one account, "As Thales was studying the stars and looking up . . . he fell into a well. A Thracian servant girl with a sense of humor . . . made fun of him for being so eager to find out what was in the sky that he was not aware of what was in front of him right at his feet."[14] But had the great Dante Alighieri (1265–1321) witnessed the incident he would not, I am certain, have made fun of Thales but would, I am still certain, have responded to the girl by saying:

> The heavens are calling you, and wheel around you,
> Displaying to you their eternal beauties,
> And still your eye is looking on the ground.[15]

Hands-on Thales responded similarly, not in words but through a practical action. "He perceived by studying the sky that there would be a good olive harvest. While it was yet winter and he had some money, he put down deposits on all the olive presses in Miletus [his hometown] and Chios [a neighboring island] for a

[13] Pliny, *Natural History* 2.53. Or see Daniel W. Graham, *The Texts of Early Greek Philosophy: The Complete Fragments and Selected Testimonies of the Major Presocratics* (Cambridge: Cambridge University Press, 2010), 25 (text 5).

[14] Plato, *Theaetetus* 174a4–8, trans. Graham, *Texts of Early Greek Philosophy*, 25 (text 7).

[15] Dante, *Divine Comedy*, trans. BookCaps (BookCaps Study Guides, 2013), Kindle Locations 10900–10901.

small sum, paying little because no one bid against him [as it was way too early for anyone to worry about the next harvest that would occur during the next autumn and winter]. When harvest time came and everyone needed the presses right away, he charged whatever he wished and made a good deal of money— thus demonstrating that it is easy for philosophers to get rich if they wish, but that is not what they care about."[16] What they do care about is the rational critique of nature.

One phenomenon that was analyzed rationally was the annual overflow of the waters of the Nile River—an unexpected phenomenon within the context of the generally dry Mediterranean summers when it begins to occur—that puzzled several Greek thinkers including Anaxagoras, Democritus, Herodotus, and Euripides. The occurrence was explained in a naturalistic way first by Thales, who ascribed seasonal northerly winds as its cause that hindered the river from emptying into the Mediterranean Sea and forced its waters to spill over its banks. Ancient Egyptians attributed the flooding to the tears of their mourning goddess Isis over the loss of her husband, Osiris. Today we know that the Nile's overflow is due to seasonal precipitation (mainly rain) on the highlands of Ethiopia (south of Egypt), where one of the sources of the Nile can be traced. Incidentally, Democritus's explanation was similar to ours.[17] That Thales was wrong is not as important as is his attempt to offer a rational explanation for this natural occurrence. Similarly, Thales and the other natural philosophers in general treated eclipses as natural phenomena, whereas the Babylonians viewed them as omens, despite the fact that the latter kept fairly accurate records for their repeated cycles. Comets, too, were generally thought of as bad omens, but not by the natural philosophers. Anaxagoras and Democritus, for example, thought that comets "are a conjunction of planets that, when coming near each other, create the illusion that they touch,"[18] an explanation, which although incorrect, is logical because it explains why a comet appears to be a strip-like light in the sky instead of point-like, as planets and stars are. Today we know that the strip-like appearance is due to a comet's tail. It is created from the sublimation of some of its ices, while a comet approaches the sun in its elliptical orbit around it.

Naturalistic interpretations of nature were the approach of all natural philosophers and remain the approach of modern scientists.

---

[16] Aristotle, *Politics* 1259a5–21, trans. Graham, *Texts of Early Greek Philosophy*, 25 (text 8).
[17] Diodorus of Sicily 1.39.1–3. Or see Graham, *Texts of Early Greek Philosophy*, 563 (text 84).
[18] Aristotle, *Meteorology* 342b25, trans. Demetris Nicolaides. Or see Graham, *Texts of Early Greek Philosophy*, 303 (text 48).

Conclusion

By reasoning that all things are ephemeral transformations of *one* primary substance of matter, Thales attempted to attribute an all-encompassing, common, and unifying principle to all the phenomena of nature, the main goal of physicists today, as well as to understand a notion of great importance in science—namely, change. The concept of change (and the degree of change) has been hotly debated for centuries. Some have accepted it as self-evident, and others have flatly denied it as an illusion. Consensus has yet to be found. Every scientist, past and present, has looked to identify a permanent principle in all of the apparent changes. What that principle might be has varied from one scientific theory to another and from one epoch to the next.

Thales was more of a practical man who accepted change undeniably. His student Anaximander was practical but also an abstract thinker. His primary substance of matter was imperceptible and although he, too, accepted change as self-evident, he also required that change in nature obeys laws and happens with measure for only then cosmic justice is preserved. But in all the conspicuous changes, he reasoned, something subtle must endure. He called it *apeiron*.

4

Cosmic Justice

Introduction

Anaximander (ca. 610–ca. 540 BCE) thought water is a bad idea for a primary substance of the universe because it's not neutral—it has an opposite, fire. And opposites destroy; they don't create one another. He taught that everything is generated from the *apeiron*:[1] a timeless, *neutral* substance, encompassing the universe and constantly transforming into competing transient opposites, but with measure to conserve the cosmic justice—without absolute dominance by either opposite. In physics, it's ubiquitous energy that's constantly transforming into ephemeral competing opposites—matter and antimatter—with measure. Curiously, however, matter ("water") is more plentiful than antimatter ("fire"). Why? Nobody knows. Where's the cosmic justice? The Higgs boson, too, is strikingly similar to the apeiron.

The Apeiron

While itself intangible, the apeiron transforms into all concrete things of every day. Thus, it is the true beginning of everything, animate and inanimate. It is also neutral, having no competing opposite. But it transmutes into opposites in struggle with one another—water versus fire, hot versus cold, wet versus dry, light versus darkness, sweet versus sour, and so on. The unjust dominance of one opposite over the other is ephemeral, for eventually it is rectified at annihilation; then, neutralized, both opposites transform again into the neutral apeiron. And since eventually the effects of one opposite cancel those of the other, their endless creations and annihilations neither add anything to the apeiron nor subtract. Thus, even through its transformations, the apeiron remains eternally conserved. In modern physics, it is energy that is conserved through its transformations into competing opposites, that of matter and antimatter, and, like the apeiron, energy is also limitless and everywhere.

---

[1] For a double etymology of *apeiron*, see Richard D. McKirahan, *Philosophy before Socrates* (Indianapolis: Hackett, 2010), 34 (Kindle ed.).

Energy and the Apeiron

In physics the notion of energy includes mass, too, since, according to Einstein's famous equation $E = mc^2$, from his theory of special relativity, energy (E) and mass (m) are equivalent and transmutable into each other—they are connected via the speed of light (c). Like the apeiron, energy is limitless, timeless, indestructible, and omnipresent even in "empty" space. Even more, energy causes change by continually transforming from one form to another (e.g., from light to heat) and from pure energy (e.g., light) into matter (e.g., electrons) and antimatter (e.g., antielectrons, also known as positrons). But even with these transformations, the total energy content of the universe is always constant. This is known as the law of conservation of energy. One can neither add more energy to the universe nor subtract any from it. Conservation laws, of which there are several in physics, ensure that changes in nature occur with measure, as in the theory of Anaximander. Measure, in modern physics, means that in all of nature's changes some things or properties remain numerically equal (e.g., energy). This equality (this measure) is in basic agreement with the view of Anaximander, who (as we'll see), to save nature and keep cosmic justice, reasoned that neither of two opposites could ever dominate totally. Now, to appreciate further the notion of measure and the similarities between energy and the apeiron, we need to first understand matter and antimatter since these, in modern physics, are opposites in struggle with each other, created from energy, and into energy once again they return.

Every particle of matter has a corresponding antiparticle of antimatter. A particle (like the negatively charged electron) and its antiparticle (the positron, which is really a positively charged electron) have the same mass and opposite electric charge (of equal magnitude). They are regarded as competing opposites since when they meet they annihilate each other by transforming completely into pure energy—like Anaximander's water and fire that neutralize each other and transform into the apeiron. Furthermore, as opposites, not only do they compete—interact via the forces of nature they obey—but, since their effects cancel each other out, they compete with measure, by obeying conservation laws such as the conservation of energy.

For example, an electron and its competing opposite, a positron, can be created out of energy, interact—they initially move apart but after a brief time in existence they recombine—and ultimately annihilate (neutralize, cancel) each other by converting their masses *entirely* back into the energy from which they came. Just like Anaximander's opposites, which are created and annihilated from and into the apeiron. Just as the apeiron remains constant during such processes, so does the energy since, according to the law of conservation of energy, the energy content of the universe is the same before the creation of the electron-positron pair, during its existence, and after its annihilation. The energy content never changes, only the forms in which it manifests itself.

Other conservation laws are also obeyed, such as that of the electric charge. In this case, the electric charge of the energy from which the pair was created was zero—pure energy always has zero electric charge: the electric charge remains zero when the pair is in existence—for an electron has an electric charge of −1 (in some units) and the positron +1—and continues to be zero when the pair is annihilated as it once again becomes pure energy.

Let us take the conservation of the electric charge one step further. Since the net electric charge is always conserved, neither type of electric charge has an absolute dominance, as Anaximander would require. Nonetheless, one type of charge has a relative dominance over the other. The negative electric charge dominates temporarily at the vicinity of the electron, whereas the positive electric charge dominates temporarily at the vicinity of the positron. It is this temporary dominance that creates the electromagnetic force of interaction and overall competition between opposite charges. It is the cause of the opposites' becoming and decaying, their attractions, repulsions, motions, conversions to and from energy, and in general, such temporary dominance is a contributing cause of the phenomena of nature.

Competing opposites are necessary for Anaximander and modern physics, as they will also be for Heraclitus, if nature is to remain diverse, eventful, and beautiful. There is an electron here but a positron there. They, and other particles and antiparticles from similar processes, convert to light and to heat. The particles form atoms, molecules, and composite objects such as the sea, the trees, the breeze, the earth, the sky, and forms of life. It is summer here but winter there. It is warm now but will be cool later; it is night now but was day earlier. The unity of the world is preserved in harmony by the very competition between opposites. Temporary dominance and the resulting struggle of opposites produce the rich plethora of diverse phenomena while simultaneously, cosmically (universally) absolute dominance is not, *should not* be allowed, for conservation laws must be obeyed. Not only does Anaximander's worldview see the nature of nature as being cosmically just, but because of conservation laws so should modern physics. Curiously, however, it appears that nature is not cosmically just, for one of the most puzzling questions of science today is why there is more matter than antimatter in the observable universe, a question to be pondered in a section later after we first elaborate on the notions of opposites and neutrality a bit more.

Not an Ordinary Thing

Anaximander reasoned that the primary substance of the universe could not have been any one of the ordinary things, such as water or fire. For they have opposition with one another, and opposites destroy; they do not generate one another. If water were the apeiron—that is, if everything in the universe were initially water—it would be impossible to have its opposite, fire, ever created, for water destroys fire; it does not generate it. And that would be terrible because in such a scenario, eventful beautiful diversity would be absent from the cosmos. Thus, something that has an opposite cannot be the primary substance of the universe for it presents a serious threat to the cosmic justice, to the unity and order of nature, to its diversity, and in fact, to the very existence of nature itself—for such type of substance, with an opposite, would cancel itself out, and thus it would cancel out nature itself!

Anaximander saved the phenomena—kept nature just, eternal, diverse, and eventful, and without the possibility of absolute dominance by any one of the opposites—by requiring that the primary substance, the apeiron, be neutral, with no competing opposite. It must be neutral to itself and to the opposites that it creates. With such choice, neither opposite is a threat to nature any longer, since their effects cancel each other out, nor is the apeiron, since it has no competing opposite to cancel out. Hence, unlike the opposites, the apeiron is permanent and indestructible. And so then is nature itself, for nature's essence is the apeiron. The neutrality of the apeiron saves the phenomena, but the opposition of the opposites beautifies them. Both neutrality and opposition are central ideas in the world outlook of Anaximander and of modern physics.

The modern physicist's version of Anaximander's reasoning would be that the presently accepted primary particles of matter, the quarks and leptons, cannot really be primary, for they have opposites—their antimatter versions—the antiparticles of antimatter, the antiquarks and antileptons, and as opposites, a particle and its antiparticle annihilate, not generate, each other. Furthermore, there are six types of quarks and six types of leptons, and the pressing question is why there are so many building blocks of matter and why do they have different general characteristics (different electric charge, mass, spin, etc.)? Why not just have one primary substance, one apeiron-like type particle?

On this issue, Nobel laureate Werner Heisenberg (1901–1976) has said, "All different elementary particles [particles that are not made of other things, thus have no substructure, such as the quarks and leptons] could be reduced to some universal substance which we may call energy or matter, but none of the different particles could be preferred to the others as being more fundamental. The latter view of course corresponds to the doctrine of Anaximander, and I am convinced that in modern physics this view is the correct one."[2]

We have seen how energy may be regarded as the apeiron, but what kind of material particle of modern physics has apeiron-like properties, including the key property of neutrality?

The Higgs Particle

The search for a universal *neutral* substance that addresses Anaximander's concern and saves the phenomena has never been more intense. The particle with the most qualities required by such a substance is the Higgs boson. It has been mathematically predicted to exist by various physicists in the 1960s, including Nobel laureate Peter Higgs (1929–), from whom it took its official name. In fact, it is required to exist in order to save the phenomena as described by the standard model of physics: the Higgs particles are thought to give mass to all material particles (such as quarks and leptons) by pulling on them, thus forcing them to slow down, clump, and form all the composite objects in the universe, from nuclei, atoms, molecules, plants, animals, planets, and stars, to galaxies, and in general, all the complexity in the universe. (The mass-giving mechanism of the Higgs is a more relevant topic for chapter 12.) With the Higgs we can explain mass, and as a result the universe is diverse, beautiful, and saved. Without the Higgs, all particles would be massless, would fly around at light speed, and would not be able to come together and form atoms or in general composite objects, including us; the universe in such a case would exist in one boring, undiversified state, which of course would be in contradiction to the actual diversified universe we live in. Indeed, then, the Higgs saves the phenomena!

Like the apeiron, the Higgs particle is intangible, neutral (in a few interesting ways, as we will explore in the next section), and the field that represents it permeates all of space. (Use of the term "field" means that something exists everywhere, whereas "particle" implies that something exists only somewhere. In quantum theory, particles are manifestations of field fluctuations. Consider this analogy: if the sea is a field, a splash in the sea—a fluctuation, an excitation—is a particle that can be detected. Because of its all-pervasiveness, the Higgs field was related to Anaximander's apeiron by Nobel laureate Leon Lederman [1922–2018].[3]) But while itself neutral, the Higgs also manifests itself as the competing opposites, as particles of matter and antiparticles of antimatter, and as some of the forces that matter and antimatter obey. In fact, it is not expected to be observed directly (analogously, neither was the apeiron); instead, its existence was confirmed indirectly by studying the behavior of various other particles that the Higgs decays into. Thus, the observed opposites in nature are in a sense different aspects of the same thing, the Higgs—or, analogously, the apeiron. Anaximander proposed the apeiron to put his opposites, and Peter Higgs proposed the Higgs field to put his opposites (the particles and antiparticles) in order to explain why they have mass.

Neutrality

To save the phenomena, neutrality must be an essential characteristic of a primary substance. The Higgs is neutral in various ways: (1) it is electrically neutral; thus, it is its own antiparticle—it is both matter and antimatter, and in this respect it has no warring opposite to be destroyed by. (2) It is also color-neutral— "color" (or more precisely, color charge) here is a property of the quarks and of the nuclear strong force (like the electric charge that is a property of the electromagnetic force) and not the color of everyday sense. (3) Moreover, its spin is zero; thus, it is direction-neutral, an unusual notion that we need to elaborate on further.

Spin (like electric charge) is an intrinsic quantum property of elementary particles. In a simplistic view, imagine the spin of a particle to be like the spin of a top around its axis. However, unlike a spinning top, which may spin slow or fast and in every direction, an elementary particle spins with a fixed magnitude and only in certain directions. Now, the direction of a particle's spin is related to the direction of its motion through space. For example, a neutrino (an electrically neutral point-like particle, belonging in the family of leptons as electrons do) is observed to always be left-handed. This means that a neutrino moves through space like a left-handed screw: it advances (moves forward) by spinning counterclockwise. An antineutrino, on the other hand, is always observed to be right-handed. It moves like a right-handed screw: it advances (moves forward) by spinning clockwise. Unlike a neutrino, an electron can be ambidextrous— or move in space by spinning in either direction. The important point is that all particles are directional—the direction of their motion through space is restricted by how they spin.

But a primary substance must itself be free of such restriction; it must be nondirectional, direction-neutral, that is, isotropic (with no preferred direction of motion)—because a left-handed substance would, in Anaximander's terms be like, say, fire, and a right-handed substance like water. If a primary substance's own motion were restricted, it would not have been able to generate the existing particles with all their various observed directionalities, which collectively are isotropic. In other words, if Anaximander's cosmic justice is to hold, a preferred (special) direction in the universe, toward which particles would be moving, should not exist. In fact, *on a grand scale*, the view of the universe is similar in all directions; thus, the universe itself is isotropic (there is no special direction in it). One piece of evidence for this is the observation of the so-called cosmic microwave background, light that comes to us from every direction in the universe, showing that the matter that emitted it had almost exactly the same temperature, about 3 degrees above absolute zero.[4] Now, since the universe is made of galaxies and stars, which in essence are made of particles, then the isotropy of the universe must really be a consequence of the isotropy of its constituent particles. And if we believe that one day we will conceive a theory of everything, describing one primary substance, such a substance must have the property of isotropy, for only then it can generate the universal isotropy that we observe today. Well, to be isotropic, direction-neutral, the spin of such a primary substance must be zero since without spin the direction of its motion cannot be restricted. The Higgs boson particle is electrically neutral as well as color-neutral, and being the only particle of the standard model with zero spin, it is direction-neutral, too! Anaximander's notion of neutrality should be a vital property of the primary substance of the universe and, consequently, of the universe itself. Nonetheless, it seems that the universe does not obey this fundamental notion of neutrality. For although it is isotropic and thus cosmically just (neutral) direction-wise, it is also unjust matter-wise; matter appears to dominate antimatter. What happened to the cosmic justice?

Why Is There More Matter Than Antimatter?

In modern cosmology, there is an open question: why is there more matter than antimatter in the observable universe? In Anaximander's terms, this problem might have been phrased: Why is there more water than fire in the observable universe? This observation makes no sense if indeed the universal substance is a kind of neutral, which transforms into equal amounts of opposites with properties that cancel each other out through the conservation laws they obey—so that the universal substance can remain neutral. Absolute dominance by any one opposite should not be allowed. Yet matter appears to have an absolute dominance in the universe. If true, and our observations are correct, where is Anaximander's

---

[4] In the Kelvin scale the absolute zero is 0 degrees Kelvin, which is −273.15 degrees Celsius, which is −459.67 degrees Fahrenheit.

cosmic justice? Are the laws of physics as we now know them really incorrect? The answer does not yet exist, but I will speculate cautiously.

Think of a creation process like that of an electron-positron pair. It obeys cosmic justice. During its existence, the region where the electron is located is dominated by matter, and the region where the positron is located is dominated by antimatter. But such dominance is relative and ephemeral. No law forbids equal and relative dominance by matter and antimatter in various regions of the universe. In fact, this would be expected if indeed equal amounts of matter and antimatter are generated by a neutral-type universal substance. What *is* forbidden is *absolute* dominance, for which the amount of either matter or antimatter in the universe is absolutely more. Now, with relative dominance in mind, we can speculate on why there is, or actually, why there *appears* to be more matter than antimatter in the observable universe.

First, the universe is immense, some 93 billion light-years across, and we haven't observed all of it yet. So an unseen part of it might be composed of mostly antimatter that could balance out the matter we see; hence, we have cosmic justice.

Second, what if our universe is not *the* universe, but rather only a mere region in it? In fact, we should not be rash in dismissing such a view, for until the early twentieth century we thought the entire universe was what we now call the Milky Way galaxy—whereas now we think there are some 170 billion galaxies. If this hypothesis holds, the puzzle of the observed asymmetry between matter and antimatter might then be resolved. Matter might be dominating temporarily in our part of the universe, but antimatter could be dominating temporarily in another part of the universe, and in such a way that neither can claim absolute dominance in the universe. Because such temporary dominance can be neutralized when such parts collide or interact, converting their matter and antimatter to pure neutral energy. Anaximander's cosmic justice would then be restored.

Cosmology

In addition to the hypothesis of the abstract apeiron, Anaximander makes another conceptual leap by holding that the earth is motionless in space and without any physical support. This happens, he argues, because earth's equal distance from everything in the sky—so it appears, anyway—is causing earth to have an equal tendency to move in every direction (e.g., equally to the right as to the left), that in turn, cancels out any potential motion and keeps earth at the center in equilibrium.[5] This is in contrast with Thales's view of an earth

---

[5] Aristotle, *On the Heavens* 295b10–16. Or see Daniel W. Graham, *The Texts of Early Greek Philosophy: The Complete Fragments and Selected Testimonies of the Major Presocratics* (Cambridge: Cambridge University Press, 2010), 59 (text 21).

floating on water and thus supported by it—for in that case, what would support the water, and what that? Philosopher Karl Popper (1902–1994) remarked, "In my opinion this idea of Anaximander's is one of the boldest, most revolutionary, and most portentous ideas in the whole history of human thought. It made possible the theories of Aristarchus and Copernicus. But the step taken by Anaximander was even more difficult and audacious than the one taken by Aristarchus and Copernicus. To envisage the earth as freely poised in mid-space, and to say 'that it remains motionless because of its equidistance or equilibrium' (as Aristotle paraphrases Anaximander), is to anticipate to some extent even Newton's idea of immaterial and invisible gravitational forces."[6] Aristarchus of Samos (310–ca. 230 BCE), sometimes called the ancient Copernicus, was the first to advance the heliocentric model, revived in the sixteenth century by Nicolaus Copernicus (1473–1543) (perhaps Copernicus could be viewed as the modern Aristarchus).

As commended by classicist John Burnet (1863–1928),[7] Anaximander's doctrine of innumerable worlds, each with its own earth, heaven, planets, stars, and especially its own (relative) center and diurnal rotation, is inconsistent with the existence of an absolute center or preferred direction of motion in the universe. With the lack of an absolute direction of motion, Burnet continues, Anaximander's argument that an earth that happened to be equidistant from everything in *its* world has no reason to move in any direction is quite sound. This is in fact a clever use of symmetry, a notion of central significance in modern physics. Symmetry, as in a circle, implies a certain constancy, a similarity that persists without change. The circle, for example, looks the same from its center at all angles, or, in Anaximander's case, the earth remains in equilibrium because of its equidistance from its own sky. Symmetry in physics does not describe just appearance. It also underlies conserved abstract properties of nature such as the conservation of energy, momentum, and electric charge. For example, conservation of energy is a consequence of the hypothesis that the laws of nature are symmetric (invariant) with respect to time translations: they work the same today as they have in the past and are expected to continue so tomorrow. Because this hypothesis has been true so far, we accept conservation of energy as a law of nature. The great mathematician Emmy Noether (1882–1935) proved mathematically various conservation laws of nature by associating them with a specific symmetry.

---

[6] Karl R. Popper, *Conjectures and Refutations: The Growth of Scientific Knowledge* (London: Routledge, 1989), 138.
[7] John Burnet, *Early Greek Philosophy* (London: A & C Black, 1920), chap. 1.

The absence of a special direction in space was employed by Democritus in describing the motion of his atoms as random (chapter 12). Though true, abandoning the notion of an absolute direction is still difficult today. We are tricked by the phenomena (such as falling objects or "the earth being under our feet and the sky up above us"), and so we often think of up above and down below as if they were really absolute up and absolute down. We don't realize that for those living on the opposite side of the earth, our relative up is really their relative down, and our relative down is really their relative up.

On Life and Evolution

Thales, Anaximander, and (as we will see in the next chapter) Anaximenes (collectively known as the Milesian philosophers since they were born in Miletus, a Greek city in Asia Minor) explained nature in terms of the variations of *one* universal substance. Nature, of course, includes us and all lifeforms. So an immediate consequence of their monistic theories is either (a) that there is no lifeless matter, but to the contrary, everything is somehow alive; each philosopher's primary substance (the water, the apeiron, or, in Anaximenes's case, the air) is somehow alive and so is everything that it transforms into. Or (b) that humans as well as all other species originated somehow from lifeless matter (the water, the apeiron, or the air).

View (a), which is known as hylozoism, was held by Thales and Anaximenes, and although highly controversial, it is still an interesting notion, for, despite 2,600 years of advancements in science and philosophy, a clear-cut distinction between animate and inanimate matter cannot be made. An unambiguous definition of what is alive or dead does not exist, as argued, for example, by four Nobel laureates: Charles Sherrington (1857–1952),[8] Erwin Schrödinger (1887–1961),[9] Werner Heisenberg,[10] and Richard Feynman (1918–1988).[11]

View (b) has a certain similarity with the premise of the modern theory of biological evolution—that regards the various species to have evolved gradually from a common ancestor (or two, possibly more) speculated to have arisen spontaneously from lifeless matter. By spontaneously, I mean that the exact mechanism of life's origin is not yet known, although chemical reactions are generally the assumed cause. Now, compared to the other Milesians, Anaximander had a more concrete and extraordinary theory of the origin and evolution of the

---

[8] Charles Sherrington, *Man on His Nature* (Cambridge: Cambridge University Press, 2009), 302.
[9] Erwin Schrödinger, *Nature and the Greeks and Science and Humanism* (Cambridge: Cambridge University Press, 1996), 66.
[10] Heisenberg, *Physics and Philosophy*, 128.
[11] Richard P. Feynman, *Six Easy Pieces* (New York: Perseus Publishing, 1963), 22.

species, including humans, that captures four specific aspects of the modern theory of biological evolution: (1) life arose spontaneously from lifeless matter, (2) more complex life did not arise spontaneously but evolved from the less complex, (3) life's adaptation to its environment, and (4) survival of the fittest.[12]

Equally important, his theory was based on an accurately analyzed *observation*. Noticing that human babies are helpless at birth and for several years thereafter, Anaximander argued that humans could not have originated with the young of the species in their present form because they would have never survived. While newborns of other animals quickly support themselves, human babies cannot survive without long parental care. Therefore, he held that humans (and in general all animals) evolved from species, precisely fish, whose newborns were more self-reliant than human (or land-animal) babies.[13]

His general doctrine was that the most primitive forms of life were generated spontaneously in the moist element as it was evaporated by the sun—note that his notion of antithesis is present here, too, as the wetness of moisture versus the dryness of the sun. These living creatures had a protective spiny membrane and were the first kind of fish. With time, he speculated, they evolved to various other forms of fish. Then, some of their descendants abandoned the liquid element and moved to dry land, adapting to different conditions and evolving to new forms of life, including humans.[14] Modern theories of biological evolution are quite similar: primitive microscopic life is speculated to have appeared spontaneously initially in water, evolved to fish, then to the sea-land transitional amphibia, to mammals, to primates, then to the first hominids from which modern humans ultimately evolved.

Newborn self-reliance was the first state in the development of life. Newborn helplessness and long parental care have developed afterward and most probably simultaneously: as a newborn need arose, an able parent addressed it. As if the evolution of a "bad" characteristic happens simultaneously with the evolution of a "good" characteristic, a notion, which if true, resonates well with Anaximander's doctrine of the simultaneous appearance of opposites: an inefficiency in some area, say, the inability to walk at birth and for several months subsequently, may evolve simultaneously with a corresponding efficiency in some other area, for example, an advanced brain, so that one may moderate the other and keep things cosmically just—for example, the evolution of an empathetic

---

[12] Aëtius 5.19.4. Or see Graham, *Texts of Early Greek Philosophy*, 63 (text 37); Censorinus 4.7. Or see Graham, *Texts of Early Greek Philosophy*, 63 (text 38); Hippolytus, *Refutation* 1.6.6. Or see G. S. Kirk, J. E. Raven, and M. Schofield, *The Presocratic Philosophers* (Cambridge: Cambridge University Press, 1983), Kindle Location 3598; Plutarch, *Symposium* 730e. Or see Graham, *Texts of Early Greek Philosophy*, 63 (text 39); Ps.- Plutarch, *Strom.* 2. See Kirk, *Presocratic Philosophers*, Kindle Location 3590.

[13] Censorinus 4.7; Hippolytus, *Refutation* 1.6.6; Plutarch, *Symposium* 730e; Ps.- Plutarch, *Strom.* 2.

[14] Aëtius 5.19.4; Hippolytus, *Refutation* 1.6.6.

brain allows parents to care for their helpless offspring for a long period. In general, the growing-up time increased with increasing brain size and complexity.

The selfless act of long parenting not only guaranteed the survival of the species but also, I believe, contributed to the overall bond between all people. Because once we began caring for our babies, we gradually began to care for our immediate and extended family, consequently increasing the chances to care more for our village, city, country, the human race, and ultimately life in general.

Conclusion

Anaximander's intellectual leap is marked by three of his theories: in cosmology of an earth motionless in space without the need of a physical support, in biology on the origin and evolution of living creatures including the human species, and certainly of his theory on the primary substance of matter. With the latter, he modeled change and diversity in terms of constant transformations of the intangible, neutral, and conserved apeiron, into the concrete, competing, and transient opposites of everyday experience, and back and forth and with measure to preserve the cosmic justice. Nonetheless, it was Anaximenes who formulated the first graspable theory of change—of how matter can transform between its various phases: the gas, the liquid, and the solid.

5

The Stepping-Stone to Truth

Introduction

In his search for the primary substance of matter, Anaximenes (who flourished ca. 545 BCE) returned to the tangible world and chose air. His way of studying nature was economical and straightforward. Starting with a single material (air) of unchangeable nature, he managed to explain the manifold of natural phenomena quantitatively, in terms of condensation and rarefaction of matter. For with these opposite processes in mind, it was no longer necessary to ascribe all sorts of different properties to each object—such as rigidity, softness, hotness, coldness, wetness, dryness, fluidity, weight, color—just how dense it was. This idea in itself has a certain truth. But from a grander point of view as regards the evolution of science, his theory was the stepping-stone to one of the most consequential truths of nature, the atom!

Condensation and Rarefaction

In Greek, "air" refers to any gas, and quite possibly in Anaximenes's view, air was vapor water. His main question, however, was how a single material, air, in its gaseous state, could be transformed into all other forms of matter and account for the overabundance of dissimilar things, while itself remaining unchanged. What mechanism or processes could be applied to air, keep its substance unchanged, yet convert air into all the different things—solids, liquids, and gases? Change, he proposed, occurs via two opposite processes: condensation and rarefaction of matter.[1] Successive condensations of gases transform them to increasingly denser matter, the liquids and solids, but successive rarefactions of solids transform them to increasingly rarefied matter and once more back to the liquids and gases, an essentially accurate idea. These processes cause changes in the density of matter but do not alter the very nature of matter (its very substance). Hence,

---

[1] Simplicius, *Physics* 24.26–25.1, Theophrastus frag. 226A. Or see Daniel W. Graham, *The Texts of Early Greek Philosophy: The Complete Fragments and Selected Testimonies of the Major Presocratics* (Cambridge: Cambridge University Press, 2010), 75 (text 3).

every object is really air—in general, made of the same material—condensed or rarefied.

Why Air?

Air is in various ways of simpler form than other everyday substances. It is highly mobile and can be found almost everywhere. It is invisible, thus apparently unstructured and symmetric, and rarefied, thus quantitatively less. Symmetry, perceptible or subtle, was and still is a much-desired characteristic for nature in both ancient and modern scientific theories. In addition, starting from less (at least in a quantitative and visual sense, e.g., rarefied invisible air) and aiming to explain more[2] (e.g., a denser, thus quantitatively more, visibly more structured, and thus in a way more complex substance), has always been the preferred approach in both science and mathematics; in mathematics, the fewer the assumptions (axioms), the more powerful a theorem is. Parenthetically, as regards religion, the reverse is true: polytheism preceded monotheism.

Now, although Anaximenes thought that fire is rarefied air and thus quantitatively less than air, still, as a primary substance fire seems to not have been adequate for him, for unlike air, fire is visible, has a variable form, and so is structured and asymmetric. Furthermore, air is needed for life through breathing, whereas fire destroys life. In fact air's traditional association with soul (from the pre-Homeric times) might have influenced Anaximenes, for he writes: "Just as our soul, being air, holds us together, so do breath and air encompass the whole world."[3] The significance of fire in the explanation of natural phenomena will be elevated in the philosophy of Heraclitus (chapter 7).

Lastly, Anaximenes was an empiricist; thus, he abstracted his theory as a consequence of careful observations of various meteorological phenomena for which air had (or so he thought, anyway) a significant role. "When it [air] is dilated so as to be rarer [more rarefied], it becomes fire; while winds, on the other hand, are condensed air. Cloud is formed from air by felting [due to condensation]; and this, still further condensed, becomes water. Water, condensed still more, turns to earth; and when condensed as much as it can be, to stones."[4] He imagined objects to be in either one distinct phase (the solid, liquid, or gas) or in a mixture of phases; "Hail is produced when water freezes in falling; snow, when there is some air imprisoned in the water."[5]

---

[2] Known as the bottom-up interpretation of nature. The top-down philosophy is introduced in chapter 10.

[3] Aëtius 1.3.4, trans. John Burnet, *Early Greek Philosophy* (London: A & C. Black, 1920), chap. 1.

[4] Hippolytus, *Refutation* 1.7, trans. Burnet, *Early Greek Philosophy*, chap. 1.

[5] Aëtius 3.3.2, trans. Burnet, *Early Greek Philosophy*, chap. 1.

From Rarefaction and Condensation to
the Atomic Theory of Matter

It will be argued that the discovery of the ancient atomic theory of Leucippus and Democritus—of atoms in the void—followed as a logical consequence of the ideas of rarefaction and condensation. And as we will see in chapter 15, modern atomic science has its roots in the atomic theory of antiquity.[6]

Softness and the Void, Rigidity and the Atoms

Softness occurs with rarefaction and rigidity with condensation, Anaximenes held, but how? Since everything is made of soft and penetrable air, why are some objects (e.g., the solids) rigid and impenetrable? Why is a piece of metal (which is supposed to be condensed air) incompressible and impenetrable, while air is compressible and penetrable? Why can we walk through air (so it seems, anyway) but not through a solid wall (which is supposed to also be air)? How do rarefaction and condensation really work, and how can we explain the varying degree of softness or rigidity in an object? Furthermore, what keeps matter together in a condensed or rarefied state?

First, let us take all four notions for granted—rarefaction, condensation, and the resulting softness and rigidity in an object. Then we ask: If we could imagine rarefaction and condensation to occur ad infinitum, what kind of an object would the absolutely most rarefied or condensed be? The most rarefied type of object would have zero density and would be *absolutely* soft, compressible, and penetrable; as if the object were void of matter; as if it were immaterial and did not exist; as if it were nothing! Now, an object void of matter is really a *void*, empty space. And so the most rarefied type of objects could be thought of as materialless gaps in space. On the other end of the limit, the most condensed type of objects would still be of the *same* substance, would have infinite density, would be *absolutely* rigid, incompressible, and impenetrable, and could be thought of as the matter that is filling up the nonempty space. These impenetrable pieces of matter that are also disconnected from each other by the void between them are precisely the atoms of Leucippus and Democritus, and the philosophically controversial void is precisely what they invented to facilitate the motion of their atoms and explain change.

---

[6] This view is also expressed in Erwin Schrödinger, *Nature and the Greeks and Science and Humanism* (Cambridge: Cambridge University Press, 1996); Werner Heisenberg, *Physics and Philosophy: The Revolution in Modern Science* (New York: Harper Torchbooks, 1962).

"There are but atoms and the void," said Democritus,[7] or equivalently, "the full"[8] and solid, and "the empty"[9] and rarefied. First note that Anaximander's opposites are present here, too, for the full and solid is the opposite of the empty and rarefied. Furthermore, "the full" seems to correspond to the aforesaid absolutely most condensed object, and "the empty" to the absolutely most rarefied object. Comparisons and details of the ancient and modern atomic theories will be carried out in chapter 12. For now it suffices to visualize tiny, indestructible (e.g., uncuttable) atoms, absolutely solid—and unable to rarefy—of all sorts of shapes moving randomly through the void, colliding with each other, either hooking with one another and clustering (condensation), or unhooking and dispersing (rarefaction). Thus, with atoms moving in the void, we understand how condensation and rarefaction are actually carried out—how matter can move, assemble, and stay together, or disassemble. And the consequent rigidity or softness is determined by the density of an object, that is, by how many atoms are cramped together within an object and by how much void is between them through which to move: the less the void, the more rigid the object; the more the void, the softer the object. Could the ancient atomic theory have been discovered through such type of analysis?

I don't know if the atomists, Leucippus and Democritus, discovered atomism by analyzing rarefaction and condensation in the two extreme limits just described, but they were certainly capable of doing so, especially the great geometer Democritus. For such type of thinking, which in mathematics is part of what is known as the theory of limits, had already been invented and applied by him in other cases (e.g., in the calculation of the volume of a cone). In fact, Democritus's knowledge on limits was commended by the great astronomer Carl Sagan (1934–1996) in this quote: "Perhaps if Democritus' work had not been almost completely destroyed, there would have been calculus by the time of Christ."[10] By "work," of course, Sagan meant, among other things, Democritus's knowledge of limits, for to invent calculus a prerequisite knowledge is the theory of limits. Calculus was finally invented independently by Newton and Gottfried Leibniz (1646–1716) in the late seventeenth century.

One other way that the mathematical analysis of rarefaction and condensation might have aided the discovery of atomism is discussed next.

---

[7] Sextus Empiricus, *Against the Professors* 7.135, trans. Schrödinger, *Nature and the Greeks*, 89.
[8] Aristotle, *Metaphysics* 985b4–20, trans. Graham, *Texts of Early Greek Philosophy*, 525 (text 10).
[9] Ibid.
[10] Carl Sagan, *Cosmos* (New York: Random House, 1980), 181.

Continuous Versus Atomic

The challenge to understand how condensation and rarefaction themselves are carried out was important for the evolution of scientific ideas because it forced Anaximenes's successors to think profoundly about the nature of matter. Consequently, they discovered two antithetical views: the continuous and the atomic (the discontinuous). The complexities associated with the former guided the mathematical genius Democritus to atomism. Schrödinger argued that the mathematical challenges of the continuum were related to similar challenges of a continuously distributed model of matter.[11] For example, we cannot tell how many points a purely mathematical line has. Analogously, if we have a material line (or in general an object), we cannot tell how many material points it has and how these points behave during rarefaction and condensation. Namely, how can an unchangeable substance of matter (e.g., Anaximenes's air), distributed *continuously* within an object, rarefy or condense? "What should recede from what [so that an object can rarefy, or what should approach what so that an object can condense]? . . . if it is a *material* line and you begin to stretch it—would not its points recede from each other and leave gaps between them? For the stretching cannot *produce* new points and the same points cannot go to cover a greater interval."[12]

In other words, can matter, modeled as continuously distributed in space, really move through other matter in order to condense or rarefy? Can new matter move into and occupy the space that is already occupied by other matter? When matter moves, where does it move into, and what does it leave behind? Why is matter able to move at all? How do condensation and rarefaction really work if matter is continuous? They do not! They work only if matter is discontinuous: made of disconnected, indivisible, and incompressible pieces—the atoms of Leucippus and Democritus—moving in the void. Rarefaction occurs when the atoms in an object recede in the empty space around them, and condensation occurs when they come closer to each other. "But by far the most important point about the rarefaction-condensation theory is that it was the stepping-stone to atomism which actually followed in its wake."[13]

Modeling matter as discontinuous (atomic) constituted the very first quantum theory, the precursor of the modern. In modern quantum theory, both matter and energy are quantized (discontinuous): matter is composed of disconnected elementary particles, the quarks and leptons, and energy comes in discrete (quantum) bundles (e.g., photons are the particles of the energy of light).

---

[11] Schrödinger, *Nature and the Greeks*, 62–65, 84–86, 157–162.
[12] Ibid., 160.
[13] Ibid., 62.

Conclusion

All challenges of rarefaction and condensation could be accounted for only through atomism (to be introduced fully in chapter 12), since such a great idea required first the development of all other great ideas conceived by Democritus's predecessors, but also in the light of mathematics (Democritus, the principal contributor of the atomic theory, was a brilliant mathematician). The significance of mathematics, not just as an abstract field of knowledge but also as a practical method to describe nature, had been realized early on, especially by the great Pythagoras as a consequence of his passion for numbers.

6

Numbers and Shapes

Introduction

Pythagoras of Samos (ca. 570–ca. 495 BCE) initiated the mathematical analysis of nature, a cornerstone practice in modern theoretical physics. "Things are numbers" is the most significant Pythagorean doctrine.[1] While its exact meaning is ambiguous, it signifies that the phenomena of nature are describable by equations and numbers—or that they are a self-organization of patterns, not chaotic or random. Therefore, nature is quantifiable and potentially knowable through the scientific method. Based on this, the underlying principle of nature is not material (e.g., water, air) but is rather a mathematical form (an equation). Since the mathematical aspect of nature is not readily realized, the doctrine was emphasizing that sense perception was merely revealing an untrue version of nature (reality), a truer version of which could be glimpsed by the intellect through modeling nature mathematically.

The Pythagoreans quantified pleasing sounds of music, right-angled triangles, even the motion of the heavenly bodies. The "Copernican revolution" (heliocentricity) is traced back to Pythagorean cosmology. But, at last, Einstein's theory of relativity shines the truth on a popular misconception related to it: that "the earth revolves around the sun (heliocentricity) is correct," and that "the sun revolves around the earth (geocentricism) is incorrect."

Plato was inspired by Pythagorean mathematics, but he replaced their "things are numbers" with things are shapes, forms, *Forms*, an abstract interpretation of nature called the theory of "Forms." The quantum-mechanical wave functions—mathematical forms that describe microscopic particles—may be considered the Platonic Forms of quarks and leptons.

The Man

Pythagoras founded a school in Croton in southern Italy, open to both men and women, where he and his students pursued various studies, including religion,

---

[1] Aristotle, *Metaphysics* 987b22. Or see Erwin Schrödinger, *Nature and the Greeks and Science and Humanism* (Cambridge: Cambridge University Press, 1996), 35.

philosophy, science, mathematics, and music. They practiced a common way of life: asceticism (through body exercise, a vow of silence, a special diet that avoided meat and fish) and secrecy (probably to keep their discoveries exclusive to their students in order to attract more members to their school). Therefore, it has always been difficult to distinguish exactly what philosophical views belong to him or to some other Pythagorean. Aristotle avoids such difficulty by often referring generally to "the Pythagoreans." Plato in *The Republic* writes specifically about Pythagoras and says that he was uniquely respected and loved by his students, not only for his knowledge but also for teaching them "the Pythagorean" way of life, best known for its high ethical standards. Wisdom, justice, and courage were among the sought-after virtues. Friendship was also highly valued. Pythagoras's dogmatic influence on his students was evident by their reference to his opinions as prophesies with the characteristic phrase "He himself said so."[2] Pythagoreanism had remained influential in Greek philosophy continuously for more than 800 years (from its birth, the end of sixth century BCE, down to third century CE).

From Earthly Harmonies to Cosmic Symphonies

Mellifluous Sounds of Music

Pythagoras's first application illustrating the role of numbers in nature was the mathematical description of mellifluous sounds of music. First, he discovered that in stringed instruments the sound of a plucked string depends on its length and tension. For example, the sound of a plucked guitar string is of a higher pitch as it is pressed down and made shorter by your finger. Then he observed that the blended sound produced by two plucked strings of the same tension is more pleasing when their lengths are in ratios of small integers—for example, 2:1 is the octave, 3:2 the fifth, 4:3 the fourth, and 5:4 the third—thus numbers forming a discrete, *quantum* set, which for the example given is the set of 1, 2, 3, 4, and 5 (obtained by arranging in sequence the numbers of the aforementioned ratios).

    The Pythagorean theory of music was a significant milestone in the evolution of science from two points of view. First, since the phenomenon of sound can be quantified, that is, it is represented by mathematical formulas, why not all phenomena? Second, if all things truly can be represented by numbers, then mathematics is the underlying and unifying principle of every natural phenomenon, even the seemingly dissimilar. With this in mind, everything may somehow be

---

[2] Diogenes Laërtius 8.46. See G. S. Kirk, J. E. Raven, and M. Schofield, *The Presocratic Philosophers* (Cambridge: Cambridge University Press, 1983), Kindle Locations 9294–9295.

related at least mathematically—in other words, a kind of master mathematical equation may be invented that can describe everything, the goal of a theory of everything. So phenomena that have no apparent relationship with one another, on a deeper level, mathematically, may prove to obey the same mathematical principle and thus have something subtle in common. We have already spoken about unification efforts undertaken nowadays in search of a theory of everything. But the first step of cosmic unification in search of a universal law was taken with the intellectually bold Pythagoreans when they connected mathematically two seemingly unrelated phenomena: their earthly harmonies with the heavenly motions. How did they do this?

The Music of the Spheres

First, they supposed that, similar to the way an object on earth moving through air can produce a sound (slow movement making a low pitch, fast movement a high one), the stars (including the sun), moon, and planets (including earth), moving through ether (the purer air believed then to fill the universe) can produce their heavenly sounds. But these sounds must blend into a song harmoniously. They reasoned that the ratios of the length of strings that produce the harmonious sounds in string instruments *must be the same* as the various ratios formed by the speeds of the revolving heavenly bodies. This requirement restricts basically the speeds and orbits of the heavenly objects to certain discrete, quantum numbers, an idea that resonates with the essence of modern quantum theory!

Relative speeds of heavenly bodies could be easily deduced by comparing each body's rising time. For example, the moon rises about 50 minutes after the stars (some reference group of them that on some day rises together with the moon), but the sun rises only about 4 minutes after the stars, well-known facts in antiquity. With this in mind, the apparent revolution speed of the stars is the fastest, of the sun the second fastest, and of the moon the slowest. Since their speeds were different, the sounds they would produce as they move through ether would also be different, to say the least. But they also had to be harmonious, the Pythagoreans conjectured, since nature was a "*cosmos*,"[3] a term credited to Pythagoras himself, a beautiful and well-ordered universe for which a cacophonous music of heavens was unaesthetic. The music of the heavenly bodies is inaudible, the Pythagoreans explained (as Aristotle tells us), because it is continuously playing and "the sound

---

[3] Aëtius 2.1.1, trans. Demetris Nicolaides. See Greek book Βας. Α. Κύρκος (Vas. A. Kyrkus), *Οι Προσωκρατικοί: Οι Μαρτυρίες και τα Αποσπάσματα τόμος Α* (*The Presocratics: Testimonies and Fragments*, vol. A) (Αθήνα: Εκδόσεις Δημ. Ν. Παπαδήμα, 2005), (Athens: Publications Dem. N. Papadima, 2005), 247.

is in our ears since our birth, thus it is indistinguishable from its opposite silence; sound and silence are distinguishable only via their mutual contrast."[4] In a parallel example, a cook does not smell his own food after a few hours of cooking it. The earthly string harmonies, which could be heard, inspired the Pythagoreans to deduce by analogy the heavenly harmonies, which could not be heard. This type of approach, to come up with a general law by analogy of something specific, is common practice in science.

Such unusual interconnection was celebrated first in 1619 with the harmonic law of Johannes Kepler (1571–1630) when the astronomer discovered that, as planets revolve around the sun in their elliptical orbits, the ratios formed by each planet's fastest speed at perihelion (a planet's closest distance from the sun) over its slowest at aphelion (its greatest distance from the sun) are very close to the Pythagorean ratios of pleasing harmonies in stringed instruments. In his book *The Harmonies of the World*, Kepler wrote, "The heavenly motions are nothing but a continuous song for several voices, to be perceived by the intellect, not by the ear."[5]

Moreover, in the beginning of the twentieth century, the seminal era of quantum theory, physicists Niels Bohr (1885–1962) and Arnold Sommerfeld (1868–1951) conceptualized the atom as a miniature solar system, with the electrons orbiting the nucleus of an atom like the planets are orbiting the sun. In their theory, however, the orbits of the electrons are quantized; they are restricted to certain discrete speeds and sizes (as were the heavenly bodies in the Pythagorean theory) that are expressible in terms of specific integers called quantum numbers that "display a greater harmonic consonance than even the stars in the Pythagorean music of the spheres [heavenly bodies]."[6] Remarkably, unlike the Pythagorean theory of planetary motion, which was quantized, the Newtonian theory was not: planets, according to Newton's theory of gravity, do not have a restriction in their speeds or orbital sizes. But they should, according to quantum theory, although their quantum behavior is negligibly small because of their large mass.

Even more so, according to the latest developments in string theory and in the words of string theorist physicist Brian Greene (1963–), "everything in the universe, from the tiniest particle to the most distant star is made from one kind of ingredient—unimaginably small vibrating strands of energy called strings. Just as the strings of a cello can give rise to a rich variety of musical notes, the tiny

---

[4] Aristotle, *On the Heavens* 290b12, trans. Demetris Nicolaides. See also Kirk, Raven, and Schofield, *Presocratic Philosophers*, Kindle Locations 9131–9133.

[5] Johannes Kepler, *The Harmonies of the World*, quoted in George N. Gibson and Ian D. Johnston, "New Themes and Audiences for the Physics of Music," *Physics Today* 55, no. 1 (January 2002): 44.

[6] Arnold Sommerfeld quoted in Gibson and Johnston, "New Themes and Audiences for the Physics of Music," *Physics Today* 55, no. 1 (January 2002): 43.

strings in string theory vibrate in a multitude of different ways making up all the constituents of nature. In other words, the universe is like a grand cosmic symphony resonating with all the various notes these tiny vibrating strands of energy can play."[7] A subtle cosmic interconnection between all things in nature, describable mathematically, was an idea envisioned by the great Pythagoras and has been consistently reaffirmed by modern physics. Mathematics nonetheless is not always rational.

The Irrationality of a Number

The proof of the Pythagorean theorem—that in a right-angled triangle the square of the hypotenuse is equal to the sum of the squares of the other two sides—was the epitome of the newly born notion of mathematical deductive reasoning, in which general theorems are proven starting from the least number of axioms. It was especially encouraging to the most important Pythagorean doctrine, "things are numbers." But it ended up also being a bad omen. For soon after the theorem's proof, its application on a special kind of right triangle—the isosceles, with its equal sides having a length of one unit—led to the discovery of a new type of number, the *irrational* number, which perplexed the Pythagoreans and shook the very foundation of their number doctrine. It was found that the length of the hypotenuse of this right triangle is equal to the square root of two, that is,

$$\sqrt{2} = 1.4142135623\ 7309504880\ 1688724209\ 6980785696\ 7187537694\ 8073$$
$$176679\ 7379907324\ 7846210703\ 8850387534\ 3276415727\ldots,$$

which does not have a precise numerical value; it can only be approximated—shown here it is truncated to 100 decimal places, for there is *literally* not enough paper in the entire universe to write such a number completely! That is, one cannot write down a precise number for the length of such a hypotenuse, only an approximate one, but we must emphasize that an approximate number is only approximate; it represents not the true length of the hypotenuse but only an approximate length. So how can things be numbers when some things cannot be assigned a precise number? To answer, we need to understand irrational numbers a bit more.

In the history of mathematics, integers (. . . , −4, −3, −2, −1, 0, 1, 2, 3, 4, . . .) were supposed to be the only numbers needed, since with their various ratios (fractions) every number that exists (including nonintegers) could, so it was

[7] *The Elegant Universe: Part 1*, PBS, October 28, 2003.

thought, be written down. For example, the positive noninteger 1/3 is expressed as a ratio of two integers, obviously 1 and 3; negative noninteger –5/4 is the ratio of the integers –5 and 4; even 0 may be thought of as the ratios 0/2 or 0/7, and so on; in fact, even integers themselves may be expressed alternatively as a ratio of two integers, for example, 8 = 16/2. Numbers that can be expressed as ratios of integers are called rational. So, for the Pythagoreans (and in general, up to that point in history), only rational numbers were thought to exist.

Since for the Pythagoreans every number was expressible as a ratio of two integers, so, too, should the length of *every* geometrical line. But they were shocked to discover that the length of the hypotenuse of the aforesaid type of isosceles right triangle could not be expressed as a ratio of two integers! That length was not a rational number. It was equal to the square root of two ($\sqrt{2}$), which turned out to be an irrational number. Irrational *literally* means that there is no ratio, none at all, that can provide an exact numerical value for $\sqrt{2}$. Hence, the $\sqrt{2}$ can only be approximated. For example, truncated to one decimal place, the $\sqrt{2}$ is equal to the number 1.4 (which, in this approximate value, can be thought of as the ratios 14/10 or 7/5); to two decimal places, the $\sqrt{2}$ is equal to 1.41 (which, in this approximation, can be thought of as the ratio 141/100). There are infinitely many irrational numbers, all numerically inexpressible by ratios. The famous number π (pi) is irrational.

The irrationality of the $\sqrt{2}$ was so shocking to the number doctrine that, according to legend, Pythagoras's student Hippasus of Metapontum (from fifth century BCE), said to have discovered it, was drowned in the deep sea in an act of divine retribution. Irrational numbers have been playing a critical role in the advancement of mathematics and physics since the time of Pythagoras. But they still present an epistemological challenge because they provide only a numerically *approximate* knowledge of nature. This "approximate knowledge" is a significant point that will be picked up again in chapter 9 in order to try to understand the fascinatingly well-reasoned but paradoxical view of Zeno, that apparent motion is not real—that, an apparently flying arrow, for example, is not really moving!

Is the Universe Arithmetical or Geometrical?

How, then, can all things be numbers if some things cannot be given an exact numerical value? They cannot if exact numbers is all that we have in mind. But they can in some broader sense. First, in general, the phenomena of nature are assigned numbers (e.g., today's temperature) determined by the various equations of modern physics. These numbers (the phenomena) are in turn compared with experimental data (again, numbers) in order to verify or falsify the hypothesis that predicts them, as required by the scientific method.

Second, we'll see in the next chapter that, within the context of the most advanced theory of matter—the quantum theory—microscopic particles lost their permanency and distinguishability and, as a result, their properties (e.g., position, velocity, energy, even their very existence) are expressed only as probabilities, only in terms of average numbers calculated from the so-called wave functions (probability functions), solutions to quantum equations. Since every macroscopic object in nature is composed of microscopic particles, then indeed all "things are numbers." Yet from another perspective, the wave functions themselves are abstract three-dimensional *geometrical* forms. Yet again, their exact *geometrical* shape depends on various quantum *numbers*. For the atoms of the periodic table of chemistry (e.g., carbon, oxygen), these quantum numbers also restrict the number of electrons that could be present in a shell (somewhere around the nucleus) in order to avoid overcrowding—for electrons (kind of like people) need their own space, an idea known as Pauli exclusion principle.

But once more, back to a geometrical representation of the universe, we'll see also in the next chapter that gravity, in Einstein's theory of general relativity, is a manifestation of the geometry of space; it is not a force as Newton had declared it.

Things then, can't be purely arithmetical; they are also geometrical. In fact, since in quantum theory the whereabouts of a *particle* are expressed probabilistically in terms of the *wave* function, then a *particle* (a *localized* entity, a probability *number*) is also a *wave* (an *expanded* entity, a *geometrical* form). This double description of nature is known as the wave-particle duality (true for both matter and light), but, equivalently, it could have been called the geometrical-arithmetical duality. Plato was the first to challenge the Pythagorean number doctrine with a geometrical theory of matter.

Plato's Theory of Forms

From Physical Forms . . .

Plato was inspired immensely by the Pythagoreans. But he thought irrational magnitudes destroyed any chance for things to be purely numbers. What then? Look all around you, what do you see? Shapes! Things are shapes for Plato. In his *Timaeus*,[8] he assumes that each form of earthly matter, earth (the solid form), water (liquids), air (gases), and fire, is composed from invisible structures of

---

[8] Plato, *Timaeus* 53 c – 57 e.

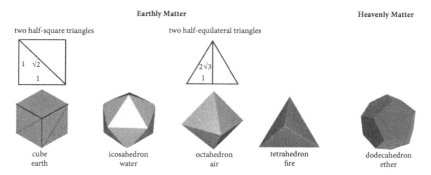

Figure 6.1 Plato's theory of earthly matter and Aristotle's theory of heavenly matter.

a unique shape (or form) that control matter's properties. If we could zoom in on, say, dirt (earth), we would notice that it is composed of microscopic cubes (which easily stack up as solid matter does)—see figure 6.1. Water is composed of icosahedra (relatively the roundest of Plato's structures so they can roll or slide as water does), air of octahedra (to flow), and fire from tetrahedra (with sharp corners to burn or cut things). Later Aristotle added a fifth essence (element), the quintessential heavenly ether—meaning "blazing" in Greek, found only in heaven[9] and thus the matter of shining stars—thought composed of dodecahedra. Our limited senses, Plato thinks, can't see these structures, only their imperfect bulk.

The earthly solids have symmetrical faces—squares the cube, equilateral triangles the other three—faces which are constructed from only two types of right-angled triangles, the half-square (having the irrational $\sqrt{2}$ as a side) and the half-equilateral (having the irrational $\sqrt{3}$ as a side). Because of their irrational side, Plato considered these triangles to be physically indivisible (atomic) and thus the basic building "blocks" of all earthly matter. From this point on in the history of science, things are a combination of arithmetic (numbers) and geometry (of physical but also abstract forms).

. . . To the Abstract Theory of Forms

With time, the idea that things have a physical unseen basic geometrical form evolved in Plato's mind into an abstract form, a noetic description of nature

---

[9] Aristotle separated heaven from the earth but we know today heaven and earth are made from the same stuff; and that, for a civilization on an exoplanet, our earth is in their heaven just as their "earth" (exoplanet) is in our heaven.

known as the theory of "Ideas" or "Forms."[10] According to it, everything we experience—physical objects of sense perception, the ocean, mountains, trees, but qualities, too, friendship, compassion, beauty—is but a mere imperfect copy, a shadow, of a greater truth: an ideal unalterable Idea, a universal Form that represents each particular object or quality. For example, an abstract Form exists that represents the ideal friendship and another Form that represents the ideal right-angled isosceles triangle. Every friendship or triangle of the sort that we have ever experienced, Plato thinks, is an imperfect copy of its corresponding true Form, which is created by God. True knowledge is achieved only with the grasp of these Forms. The quantum mechanical wave functions (which are mathematical forms) may be considered the Platonic Forms of microscopic particles.[11]

The Value of Mathematics

The theory of Forms may have its origin in Pythagorean mathematics. Plato himself was an accomplished mathematician. The entrance to his Academy had the inscription: "Nobody Untrained in Geometry May Enter My House."[12] Pythagoras showed us that mathematics quantifies nature and that it is the language of science—we can measure nature in order to verify or falsify our scientific hypotheses. And even when the law of a natural phenomenon has not yet been discovered, the law is assumed to exist, as is the mathematical equation that can express it. This, in fact, is the very premise of science. Without such an attitude, science cannot be done and truth cannot be found.

Geocentric Versus Heliocentric: The Relative Truth

Pythagorean Cosmology

Being a great geometer who understood well the relationships of spheres, flat surfaces, and lines, Pythagoras was probably the first to deduce that the earth is spherical. Several observations might have aided him in reaching such a conclusion. During a lunar eclipse, the shadow of the earth on the moon is a circular

---

[10] Plato, *Republic*, Book V to Book VII; Bertrand Russell, *The History of Western Philosophy* (New York: Simon & Schuster, 1945), 119–132; Karl R. Popper, *Conjectures and Refutations: The Growth of Scientific Knowledge* (London: Routledge, 1989), 90–96.

[11] Schrödinger, *Nature and the Greeks*, 122; Werner Heisenberg, *Physics and Philosophy: The Revolution in Modern Science* (New York: Harper Torchbooks, 1962), 45–46.

[12] Popper, *Conjectures and Refutations*, 87.

arc. The masts of receding ships disappear last (and, equivalently, appear first when ships are approaching). Pythagoras himself knew that the evening and morning "stars" are really the planet Venus.

But the most notable achievement of the Pythagoreans in cosmology, often credited to the Pythagorean Philolaus (ca. 470–ca. 385 BCE), was when they displaced the earth from the center of the universe and imagined it in motion. So the earth revolves around a center occupied by fire, called Central Fire, and so do the moon, sun, planets (Mercury, Venus, Mars, Jupiter, Saturn—the ones known to antiquity, as only these are visible without a telescope), and the fixed stars— termed so because of their apparently fixed position with respect to one another; on the other hand, *planet* means literally "wanderer" because planets were changing their position among the fixed stars.[13] Central Fire is invisible because the inhabitable hemisphere of the earth faces always away from it, whereas the side of earth that always faces it is uninhabitable because it is too hot. Incidentally the moon's synchronous motion—according to which its rotational period around its axis is the same with its revolution period around the earth—produces the same effect: the near side of the moon always faces the earth, whereas its far side always faces away from earth, making it always invisible to an earthbound observer. Revolving around Central Fire is another body, the anti-earth, termed so because of its position.[14] It is imagined to always be in the same direction as the uninhabitable hemisphere of earth, so, like Central Fire, it, too, is invisible. It is not certain why anti-earth was required (some scholars speculate that it was needed to explain eclipses), or even whether anti-earth was really a planet at all— for due to its position, anti-earth might have simply been the uninhabitable hemisphere of earth.

In addition to its revolution around Central Fire, earth also rotates around its own axis daily, accounting for the apparent revolution of the sky. This understanding was in audacious opposition to the popular view of an immobile earth at the center of the universe as well as to the evidence of the senses that do not feel earth's motion. In an analogy, to understand the apparent revolution of the sky, pretend to be the earth and stand at the center of a room. Then begin to rotate around the axis of your body, say, counterclockwise. The walls, which you can think of as the sky (with the sun and stars), appear to revolve around you in the reverse direction, clockwise.

Only Central Fire is self-luminous; all other bodies are shining with reflected light from it. In fact, this might be the justification of its postulated existence: since the moon is shining with reflected light—an ancient knowledge—and by the cycle of day and night, so, obviously, is the earth. But why not the sun,

---

[13] Schrödinger, *Nature and the Greeks*, 45.
[14] Ibid.

which in many ways is like the moon—in motion, shape, size, eclipses, color—and all heavenly bodies? If yes, a source of light had to be speculated, hence the Central Fire.

That the Pythagorean system is neither geocentric nor heliocentric is actually a quite justifiable cosmological theory. Since every visible celestial body appears to be moving, perhaps so should the earth, the Pythagoreans might have thought. Now the difficulty with imagining the earth in motion has its origin in our deceptive senses, namely our eyes. Our inability to detect the actual distance of what we see, in particular the stars, is tricking us into thinking that all bodies are the same distance from us. Therefore, being also in different directions, they appear fixed on a hemispherical dome, which is part of what we call the sky. As the sky appears to revolve daily around us, the new stars brought into view appear, for the same reason, to also have the same distance from us as all the rest. Thus, we imagine every star fixed on a spherical sky—even though at any one time we see only a hemispherical sky—with us on earth at its center. In addition to that, the apparent daily revolution of the dome-like sky around us is easily tricking us into thinking that the earth is absolutely motionless at the sky's center and therefore is at the absolute center of the universe, as if the earth occupies a special position in the universe. This is, in fact, the geocentric view, which, due to our imperfect senses and our initially uninformed intellect, naturally emerged as the first cosmological model. But the Pythagoreans were well aware of the unreliability of the senses, and they were also accomplished mathematicians with sharp, critical minds. Moreover, being people of virtues, such as humbleness, the Pythagoreans had no difficulty displacing the earth and themselves from the center and purpose of the universe.

No Special Center

Influenced by the Pythagorean cosmology, philosopher Heraclides Ponticus (ca. 390–ca. 310 BCE) proceeded to devise his own. He explained correctly the varying brightness of Mercury and Venus as the result of their varying distances from earth. The additional observations that these planets seem to always follow the sun—when visible, each of them (but independently of the other) either rises just before the sun or sets just after the sun does, especially so Mercury because it is closer to the sun—prompted him to imagine these two planets revolving around the sun, thus justifying their varying distance from earth and consequently their varying brightness, and the sun revolving around an immovable earth at the center of the universe. This partial heliocentric view—with only two planets revolving directly around the sun—became a full-blown heliocentric theory when Aristarchus proposed that all planets including earth revolve around the sun.[15] While this theory was rejected in favor of the geocentric view, it was revived much later by Copernicus.

Traced back to Pythagorean cosmology are the first steps away from the prejudices of the geocentric and anthropocentric worldview and the inspiration for the discovery of the heliocentric worldview. However, perhaps due to scientific misrepresentation of the topic, the popular perception is that the heliocentric model is correct and the geocentric model incorrect. But the profundity of the heliocentric model is really this: (1) it is *another* point of view *as good as the geocentric*—though initially, like the geocentric, it, too, was incorrectly perceived as absolute, as if the sun were the absolute center of the universe (as Copernicus held in his *On the Revolutions of the Heavenly Spheres*, who inspired Galileo Galilei [1564–1642] to hold the same hypothesis in his book the *Dialogue Concerning the Two Chief World Systems*)—and (2) since another center is as good as the previous one, the notion of an absolute center of the universe is abolished. In fact, in modern physics the any-center view is correct. A particular center is chosen merely for its conceptual and mathematical convenience for the understanding of a physical phenomenon and is not to be misinterpreted as absolute or uniquely correct. This view is supported by special relativity (see next subsection). It is also supported by astronomical observations, including the discovery of numerous new galaxies, each with billions of stars revolving relative to the galaxy's center, and generally observations indicating that the universe is isotropic (thus, no one location is more special than another). Finally, such a view may be accepted based merely on pure humility, that neither the earth nor the sun should occupy a special center, and in general that no point in the universe should be more centered or privileged than another. The universe has neither an edge nor a center, and the laws of physics apply equally the same everywhere. "The merit of the Copernican hypothesis [that (1) annually *earth* revolves around the sun, not the sun around the earth and (2) diurnally *earth* rotates on its axis, not the sky around earth] is not *truth*, but simplicity; in view of the relativity of motion [from Einstein's theory of relativity], no question of truth is involved."[16] Equally correct (as will be emphasized a bit more later in this section), we could imagine that (1) annually either the earth revolves around the sun or the sun around earth and (2) diurnally either earth rotates on its axis, or the sky revolves around earth. Space and time were absolute in Newtonian physics but became relative in Einstein's theory of special relativity. This means that for relativity an *absolute* frame of reference—a special location of observation that can be used to refer to absolute motion—does not exist. There exist only *relative* frames of reference that can be used to refer to relative motion. Hence,

---

[15] Russell, *History of Western Philosophy*, 214.
[16] Ibid., 217.

we can choose any center *relative* to which something can be at rest or in motion. But a special center for absolute rest or absolute motion is utterly meaningless. Space, time, and motion are all relative. Let us elaborate.

Newtonian Absoluteness Versus Einsteinian Relativism

Newtonian Physics

In Newtonian physics, space and time are absolute and thus independent of an observer's relative motion. This means that space distances and time intervals are unchanged by motion. For example, the length and mass of an object are the same for all observers independently of their location or motion relative to the object or relative to one another. The same for them is also the way time passes. Twins, for instance, have always the same age with respect to one another, whether they move or not relative to each other. In general, two events that are simultaneous for one observer are simultaneous for every observer—absolute simultaneity. Space is a kind of preexisting passive (unaffected, in a sense "disconnected" from everything else) immutable playground where objects exist and events occur while time flows steadily in the background as if a cosmic clock existed that showed the same exact time for everyone and every location in the universe. "Now" and "here" are absolute concepts for Newton—that is, everyone agrees when and where something happens. But Einstein's theory of special relativity proved all these to be false, in spite of the fact that all these are how we experience the world daily.

Special Relativity

In special relativity the speed of light in a vacuum, designated c, is always 671 million miles per hour—put differently, in 1 second light travels as far away as is the distance of eight times around the earth. It is the same in all reference frames (for all observers, moving or not relative to the light source). It is also a kind of cosmic speed limit—*absolutely constant*—for although it could be approached, absolutely no material object can travel as fast as or faster than light. This fact has nothing to do with engineering. It is not because we don't have high-powered engines to accelerate an object to the speed of light; rather, this fact is how nature behaves. It is a law of nature that has withstood the scrutiny of experiments since 1905, the year it was postulated by Einstein in his theory of special relativity. If c were not a cosmic constant, causality would be violated and the universe would be paradoxical: a message, travelling faster than c, could be sent to my past to prevent my parents (my cause) from meeting. But how could I (the effect) then exist with my cause eliminated? That I exist is evidence c must be a cosmic constant, but the consequences of that are radical.

Two of the most dramatic consequences concern space and time: they are no longer absolute, they are relative—dependent on an observer's relative motion. They are combined mathematically (by the so-called Lorentz transformation) into a continuum called space-time. Space distances and time intervals do change with respect to an observer's relative motion. Relative space means that a moving object contracts in the direction of motion, as seen by (relative to) a stationary observer—a phenomenon known as length contraction. Relative time means no cosmic clock, that the passage of time in a moving clock (say, aboard a moving spaceship) is dilated; it is slower relative to the passage of time in a stationary clock on earth—a phenomenon known as time dilation. To function properly, the Global Positioning System (GPS), a common cellphone app, takes time dilation into consideration. If it didn't, the GPS receiver in your car or phone would miss your destination. Parenthetically, the GPS must also take into account another time effect, predicted by general relativity. That clocks in orbit, where gravity is weaker compared to the ground, run faster relative to clocks on earth.

Interestingly, time dilation makes time travel possible because using it we can travel into the future. Suppose Earthly and Heavenly are twins. Heavenly likes to journey in space, while Earthly prefers to stay on earth. If Heavenly travels at a speed close to c, upon her return to earth she will realize that she has aged less than Earthly (and all the other people or things on earth). How much less depends on the duration of her trip as well as how close her speed was to c. For example, if her speed was 99.5 percent of the speed of light, then for every 1 year that Heavenly ages during her trip, Earthly ages 10 years. But if her speed was 99.99 percent of the speed of light, then for every 1 year that Heavenly ages, Earthly ages 71 years. So a twin who takes a trip into space will age less with respect to the twin on earth. Here we emphasize that Heavenly feels no different, as regards the passage of time, while traveling. The difference in age is noticed when the twins compare notes, for example, meet again.

In general, if you travel at a speed close to the speed of light, the time elapsed for you will be less compared (relatively) to the time elapsed for those not traveling with you. Hence, because of time dilation, you can then travel into the future of those not taking the trip with you; like the astronauts in the 1968 film *Planet of the Apes*, who aged only 18 months during their near-light-speed journey and returned to find a postapocalyptic earth where the elapsed time since they left was 2,006 years. So by taking a trip at high speeds you may return to earth at some future century of your choice. You may enjoy great developments of a more advanced civilization in that future century. The downside is that, if it is too far into the future, none of your familiar people may be alive to welcome you. Would such a trip be worth taking?

There is yet another fascinating consequence of the constancy of the speed of light. It allows us to see the past. In fact, we do it all the time. Looking out in space is looking back in time. And the further out we look, the further into the past we see. This is so because starlight takes time to travel from the distant stars to our eyes. The speed of light is not infinite, so light messages are not transferred instantly. The light from the sun, for example, takes about 8 minutes to reach our eyes. This means that observing the sun at, say, 12:00 noon is actually seeing how the sun was at 11:52 a.m. But Polaris, the North Star, is about 434 light-years away from us (i.e., its light takes 434 years to reach us). So looking at Polaris tonight is actually seeing how Polaris looked 434 years ago. Polaris may not even be there now!

There are still other effects of special relativity. A moving object becomes more massive relatively to its mass at rest. As such an object is approaching c, its mass is approaching infinity, but an infinite mass requires infinite force—which doesn't exist—to increase its speed to precisely c; thus, c is unattainable by material objects. Also events that are simultaneous for one observer (say two babies are born at the same time) are not so for another observer in motion relative to the first observer (one baby is born before the other). All these so-called relativistic effects become evident only at high speeds, however, those comparable to the speed of light. Because the everyday phenomena involve speeds so much smaller than the speed of light, we are tricked into thinking that Newtonian physics is true. It is nevertheless an excellent approximation of truth, for at the limit of low speeds the equations of special relativity reduce to the Newtonian ones. The equivalence of mass and energy (expressed by the equation $E = mc^2$), length contraction, time dilation, and the relativity of simultaneity, are some of the most startling consequences of special relativity.

Special relativity, published in 1905, deals with the special case of uniform motion (of constant speed in a straight line). General relativity, published in 1916, deals with the general case of accelerated motion (of changing speed and/or direction). Incidentally, since falling objects fall by accelerating, general relativity is really a theory of gravity, more advanced (and with a different vision, philosophy) than Newtonian gravity. Now, space, time, and motion are relative in both special and general relativity. Relative motion means that motion can be discerned or measured only relatively to a reference frame (or an observer). "Every motion [uniform and accelerating] must only be considered as a relative motion,"[17] said Einstein. By contrast, Newton thought for example that rotational motion is absolute.

---

[17] Albert Einstein, *Relativity: The Special and General Theory* [New Kindle Edition with Readable Equations] (Kindle Locations 615–616).

The Relative Motions of the Earth, Sun, and Sky

Hence, in light of special and general relativity, for which space, time, and motion are relative, as regards annual motion, according to the geocentric model, relative to the earth (that is, relative to an earthbound observer), the sun appears to revolve around the earth in 1 year. But equally correct, according to the heliocentric model, relative to the sun (that is, relative to a hypothetical sun-bound observer), it is the earth that appears to revolve around the sun in 1 year. Likewise, as regards diurnal motion, the sky (with the sun and stars) appears to revolve westward relative to the earth daily (and so the sun appears to rise from the east and set in the west relative to the earth daily)—recall that in our earlier analogy the walls appear to revolve clockwise relative to you. But equally correct, the earth appears to rotate eastwardly on its axis relative to the sky daily—you appear to rotate counterclockwise on your axis relative to the walls. This difference (of what is moving with respect to what) "is purely verbal; it is no more than the difference between 'John is the father of James' and 'James is the son of John.'"[18] The view of the daily eastward rotation of the earth (relative to the sun and all other stars in the sky) is merely more economical (for only one object, earth, is in relative motion) than the view of the daily westward revolution of the sun and of the myriad stars of the sky (relative to the earth).

The genius of relativity, writes, "Take two bodies, the sun and the earth, for instance. The motion we observe is again *relative*. It can be described by connecting [considering] the c.s. [coordinate system, relative center] with either the earth or the sun. From this point of view, Copernicus' great achievement lies in transferring the c.s. from the earth to the sun. But as motion is relative and any frame of reference [center] can be used, there seems to be no reason for favoring one c.s. [relative center] rather than the other."[19]

Hicetas (ca. 400–335 BCE), a Pythagorean, realized the relativity of motion before Galileo, Bishop Berkeley (1685–1753), Ernst Mach (1838–1916), and Einstein: "The Syracusan Hicetas, according to Theophrastus, believes that the sky, the sun, the moon, the stars and generally all things that lie above are motionless and that nothing in the cosmos moves, except earth; as it rotates around its axis with high speed, it causes the same effect that would have been caused if earth were at rest and the sky moved [rotated]."[20] Hicetas's student, Ecphantus (ca. fourth century BCE), adds another important detail, "like a wheel, the earth

[18] Russell, *The History of Western Philosophy*, 540; Isaac Asimov, *Understanding Physics* (Dorset Press, 1988), vol. II, 117–119.
[19] Albert Einstein and Leopold Infeld, *The Evolution of Physics* (London: Cambridge University Press, 1938), 222–223.
[20] KIK. Acad. Pr. II 39, 123, trans. Demetris Nicolaides. See Greek book *Προσωκρατικοί (Presocratics)*, vol. 13 (Athens, Greece: Kaktos, 2000), 73, https://www.kaktos.gr/001110 (accessed July 14, 2019).

rotates around its axis from west to east."[21] Of course, everyone (then and now) knows that the sun, moon, and stars (the sky in general) appear to rotate from *east* to west. This implies that Hicetas and his student understood yet another aspect of the relativity of motion: not only that things can be viewed in motion with respect to one another but also that their relative motions are in opposite directions. Hicetas was cited by Copernicus in his *On the Revolutions of the Heavenly Spheres*. Newton, too, in his *The System of the World* "attributes to antiquity (correctly) . . . the Copernican revolution: 'It was the opinion of the ancient philosophers that in the highest parts of the world the stars remain fixed and motionless, and that the Earth turns around the Sun.'"[22]

Mach's Principle

In criticizing the absoluteness of space, time, and motion of Newtonian physics, Mach, like Berkeley before him, held that (1) all motion is relative.[23] Mach, however, accepted one of the tenets of Newton's law of universal gravitation, that (2) an object's motion is affected by the gravity it experiences from all other objects in the universe *instantly*. Einstein coined ideas (1) and (2) as Mach's principle and found them inspiring in the development of his relativity.[24] But in the end, Einstein embraced only (1).[25] He rejected (2), because in relativity the fastest that communication is postulated to travel is only at the speed of light, not instantly. (But don't dismiss instant interaction just yet; wait until the section on "Quantum Entanglement" in chapter 8.)

I find curiously interesting the following similarity between Plato's explanation of motion and Mach's idea (2). Both Plato and Mach rejected atoms and the void, and accepted the plenum (that *all* space contains matter). Plato explained motion as the rotation of a wheel. Since there is no void, he thought everything is connected (either via direct contact or via matter in-between). Thus, an object can move by pushing its surrounding matter—which is in direct contact with it—while, simultaneously, that matter is pushing its own surrounding matter (and that matter its own), causing everything in the universe to instantly interchange position and direction and move much "like a rotating wheel"[26] (where

---

[21] Aëtius 3.13.3, trans. Demetris Nicolaides. See Greek book *Προσωκρατικοί (Presocratics)*, vol. 13 (Athens, Greece: Kaktos, 2000), 73, https://www.kaktos.gr/001110 (accessed July 14, 2019).
[22] Isaac Newton quoted in Carlo Rovelli, *Reality Is Not What It Seems* (New York: RiverHead Books, 2017), 72 (Kindle ed.).
[23] Popper, *Conjectures and Refutations*, 169, 171–172.
[24] Ibid., 172.
[25] Albert Einstein, *Relativity: The Special and General Theory* [New Kindle Edition with Readable Equations] (Kindle Location 751).
[26] Plato, *Timaeus* 79 c; see also 80 c, 79 b, 52 e, 57 d.

all points/matter on it move/s simultaneously). Hence, every time something is moving, everything in the universe is simultaneously moving with it—for example, if I spin about my body's axis, everything in the universe spins simultaneously with me, too. Motion is communicated instantly to all objects however far they are, Plato implies, as is Mach's idea (2). Pythagoreans certainly influenced Plato, who plausibly influenced Mach, who certainly influenced Einstein.

Epicurus (chapter 13), however, who accepted atoms and the void as a theory of matter, derided Plato's explanation. There are those who "maintain that water yields . . . to the . . . fish that push against it, because they leave spaces behind them into which the yielding water can flow together. In the same way, they [Plato, Aristotle and others who taught the plenum and rejected the void as a means of motion] suppose, other things can move by mutually changing places [as in the analogy of a rotating wheel], although every place remains filled. [But] how can the fish advance till the water has given way? And how can the water retire when the fish cannot move? . . . [The alternative:] . . . things contain an admixture of vacuity whereby each is enabled to make the first move."[27]

Conclusion

Within the context of modern physics "things are numbers" but also abstract forms indeed. What's more, many apparently unrelated things (phenomena) have already been unified; they are found to obey the same fundamental mathematical equation and thus the same natural law (e.g., the electroweak unification). These findings point clearly to the subtle cosmic interconnection (of mathematical nature) anticipated by the Pythagoreans. But in addition to the aid of mathematics, to find the Logos (reason) of such inconspicuous connections, one needs to be unconventional, to be able to unite diverse fields of knowledge, and to focus a keen eye on the elusive. For only then may one unveil the common characteristics that different phenomena have in all of nature's changes, the perceptible but also the discreet.

---

[27] Lucretius, *On the Nature of the Universe* 1.373–384, trans. R. E. Latham (London: Penguin Books, 2005), 19.

7

The Changing Universe

Introduction

Everything is constantly changing, and nothing is ever the same, Heraclitus of Ephesus (ca. 540–ca. 480 BCE) proposed, and in accordance with Logos, the intelligible eternal law of nature. Thus, everything is in a state of becoming (in the process of forming into something) instead of being (reaching or already being in an established final state beyond which no more change will take place). This means that things, *permanent* things, no longer exist—for they contradict his theory of constant change—only events and processes exist. His doctrine has found strong confirmation in modern physics, for, according to it, absolute restfulness and inactivity are impossibilities. Points in Einstein's four-dimensional space-time continuum are events, and so are the quarks and leptons—for, unlike in deterministic Newtonian physics, matter in probabilistic quantum physics lost its permanence and identity because of the Heisenberg uncertainty principle. Moreover, all happenings, evidence suggests, are consistent with a single universal law.

Strife and Harmony

For Heraclitus everything in nature is characterized by opposites that are struggling. "We must recognize that war [the competition between opposites] is common, strife is justice, and that all things happen according to strife and necessity."[1] So without strife, as Homer had wished, the universe would be led to its destruction because events and processes could not have existed without some force that promotes change: "Heraclitus criticizes the poet who said, 'would that strife might perish from among gods and men' [Homer, *Iliad* 18.107]; for there would not be harmony without high and low notes, not living things without female and male, which are contraries."[2] Hence, "strife is justice" because change,

---

[1] Origen, *Against Celsus* 6.42, trans. Graham, *Texts of Early Greek Philosophy*, 157 (text 58).
[2] Aristotle, *Eudemian Ethics* 1235a25–29, trans. Graham, *Texts of Early Greek Philosophy*, 157 (text 60).

for Heraclitus, is caused by the strife of the opposites. Without strife, change would not occur.

Now, like Anaximander, Heraclitus, too, requires cosmic justice by such strife. In fact, he argues that not only is absolute dominance not allowed by any of the opposites but quite the reverse, that harmony is born from their strife. "Attunement [harmony] of opposite tensions, like that of the bow and the lyre."[3] This harmony of strife is the result of a subtle underlying unity shared by the opposites; generally, it is the result of the common characteristics that different things have. For example, the property of mass is common to both the different objects the earth and the sun. As a result (according to Newton's third law discussed later), each body attracts the other with the same strength! Discovering and understanding such unity is understanding Logos, but to manage this is difficult because "nature loves to hide."[4] Nonetheless, Newton and several scientists thereafter have managed it.

Action-Reaction

Newton's action-reaction law, his third law of motion, describes the strife of opposite forces but also their subtle unity and harmony. According to it, for every action force there is an equal reaction force in the opposite direction. For example, the force exerted on a nail by a hammer has the same strength and is in opposite direction to the force exerted on the hammer by the nail. The competing opposites are the competing forces acting in *opposite* directions, but they do so with *equal* strength—so their unity in strength is mathematically expressible, in other words, action force = reaction force. A force (the action) cannot exist by itself; it exists only in relation to its opposite (the reaction)—thus, Homer's wish to eliminate strife is unrealizable in Newtonian physics. In fact, generally in physics "physical action always is *inter*-action, it always is mutual."[5]

Analogously, in Newtonian gravity, the earth is attracting you *downwardly* with the *same exact* force as you are attracting the earth *upwardly*! Your weight is the strength of this mutual force. The earth and sun attract each other with forces of *equal strength* (unbelievable but true) and *opposite* directions, and as a result both celestial bodies move harmoniously through space and time. Both bow and lyre obey Newton's third law, too. In a bow, the cord is pulling each of

[3] Hippolytus, *Refutation* 9.9.2. Or see Bertrand Russell, *The History of Western Philosophy* (New York: Simon & Schuster, 1945), 43.

[4] Themestius, *Orations* 5.69b, trans. Demetris Nicolaides. Or see Graham, *Texts of Early Greek Philosophy*, 161 (text 75).

[5] Erwin Schrödinger, *Nature and the Greeks and Science and Humanism* (Cambridge: Cambridge University Press, 1996), 157.

the two limps (the flexible upper and lower parts of a bow) in one direction—the action force, which is along the cord and toward its midpoint. Whereas the limps respond by pulling the cord in the opposite direction—the reaction force, which is also along the cord but away from its midpoint. Furthermore, action and reaction are forces of equal strength. So the bow's apparent rest is really the result of the constant strife between opposite equal tensions, the action and re-action. Newton's third law applies even while the cord is being drawn in order to shoot an arrow (or as the cord is being released and shooting the arrow); or as the strings of lyre are at rest, or as they are plucked, producing their sweet notes of music. Even more impressive is that the apparent inactivity, at the macroscopic level, of the bow or lyre at rest (or any other object), is, at the microscopic level, really a frantic and endless activity of particle exchange; for force, on the micro-scopic level, is really an eventful process. And constant change, even the imper-ceptible, is indeed a fact.

Force in Quantum Theory

According to the standard model of quantum theory, the forces of attraction or repulsion between the particles of matter (the quarks and leptons) are caused by the constant exchange of particles of force—called force-carrying particles or messenger particles since they carry the message of the force. The exchange of force particles transfers energy between the particles of matter, causing a change in their own energy, speed, and direction of motion and making them attract or repel.

The massless photons mediate the electromagnetic force; the massless gluons transfer the nuclear strong force (gluons, "glue," bind the quarks to form protons and neutrons, for example); the massive W^+, W^-, and Z^0 particles (of positive, negative, and zero electric charge, respectively), the nuclear weak force; and the massless gravitons are speculated to mediate gravity.[6] As regards the gravitons, a complete theory that describes them has yet to be discovered, and, equally im-portant, no experiment so far has confirmed their existence.

The electric repulsive force between two electrons, for instance, is medi-ated by the continual exchange of photons that, traveling at the speed of light, are emitted and absorbed by the electrons. Namely, one electron rebounds by emitting a messenger photon, and the other electron rebounds by absorbing the photon. Repeated processes of this kind mean that the exchanged photons knock the interacting electrons further and further apart. It is this continual exchange

---

[6] Leon Lederman and Dick Teresi, *The God Particle: If the Universe Is the Answer, What Is the Question?* (Boston: Houghton Mifflin, 1993).

of photons that manifests itself as the electric repulsive force between the two electrons. Similar processes can explain the other forces.

Through the continual exchange of the particles of force, the particles of matter move nonstop and combine with one another to form atomic nuclei, atoms, molecules, and composite objects like bows and lyres. Thus, even an apparent static equilibrium of an object at the macroscopic level, down to the microscopic level, is really an eventful, complex, and endless process of particle exchange. Nature is constantly changing.

Logos

Newton's third law of motion or the more detailed description of a force by the standard model may be viewed as part of Logos. In the third law, the underlying unity is the equality of the strength of the opposite forces. In the microscopic interpretation of force, unity is expressed by the conservation laws obeyed by the particles through their interactions (strife); that is, as the particles of matter collide with the particles of force, their net energy (or momentum, to name just two properties that are conserved) before collision *equals* their net energy (or momentum) after collision—again, the Heraclitean unity between competing opposites is expressible mathematically in physics; that is, energy before = energy after, or, momentum before = momentum after. Of course, the actual equations are more descriptive, detailed, and written with mathematical symbols.

Also, matter and antimatter are opposites in strife. Their Logos are the various laws they obey, including gravity (of Newton or of Einstein, his general relativity), electromagnetism (of Maxwell or quantum theory's), the standard model, string theory, loop quantum gravity, and so on. And the underlying unity consists of the various conservation laws with which each process involving matter and antimatter must comply. The resulting harmony in the strife of matter and antimatter is the general organization of the world (a notion to be revisited in the section "Organization").

In modern physics we are striving to understand various phenomena, first by isolating them and finding which laws they obey. But as in Heraclitean philosophy, according to which true understanding is achieved by identifying common characteristics that different things have, the real picture emerges only when we manage to connect our understanding of isolated and seemingly different phenomena and discover the bigger truth, the Logos they all obey. In modern physics one of the key scientific principles, which is part of Logos, is the Heisenberg uncertainty principle. It will help us understand the doctrine of Heraclitean change from within the context of quantum theory.

The Uncertainty Principle

The most consequential, mind-boggling law of quantum theory—its very heart and soul—is the Heisenberg uncertainty principle. This principle discusses how nature limits our ability to make exact measurements regardless of how smart or patient we are or the sophistication of our experimental apparatus. Namely, as a consequence of the very act of observation, the observer always disturbs the object being observed a certain minimum way, causing the result of a measurement to be uncertain. We can measure very accurately the position and velocity of a large-mass object, such as a car or a planet, without significantly disturbing it. We can watch it move and even predict its path of motion. But if instead we had a small-mass object, such as a microscopic particle—an electron, a proton, an atom, even a molecule—we could not measure exactly both its position and its velocity; nor could we observe it in a path of motion or predict its path. Quantum theory can describe only where (say) an electron is likely to be, not where it was, is, or will be. Before the uncertainty principle was discovered, absolute accuracy in a measurement, at least in theory, was considered axiomatic, but not anymore.

Suppose we want to observe an electron, hoping to "see" where it is and determine how fast it is moving. To do so, we, the observer, must shine a light of a certain wavelength ("color") upon it—bounce a photon off it. The light (the photon), which is scattered by the electron, will then enter our microscope, be focused, and be seen by our eye. It is the scattered photon that we actually see in an act of observation. Now to illustrate how the observation itself creates the uncertainty in a measurement, we discuss such an act in two steps. Step one discusses what happens to the electron when light is shined upon it—when the photon collides with it. Step two discusses how clearly the electron can be seen (focused) through the microscope. It is the combination of the effects from these two steps that produces the celebrated uncertainty principle.

Step One: The Collision

As a result of their collision, the bouncing photon transfers some of its energy (and momentum) to the electron and disturbs it (much like when one billiard ball disturbs the motion of another when they collide). But there is no law that can determine the amount of energy imparted on the electron by the photon. Thus, the photon pushes and disturbs and changes the velocity of the electron unpredictably. This means that the electron may have a range of possible recoil velocities; hence, its velocity cannot be known precisely: there is an uncertainty in its velocity. On the other hand, the disturbance introduced by a photon bounced off a car or a planet is undetectably small because, compared to an electron, the mass of a car or a planet is huge; just think how much more difficult it is for anyone to push and disturb a real car, which weighs a lot, compared to pushing a toy car, which does not weigh much. The velocity and location of a car or planet can be measured almost with absolute precision. This is in fact another reason that classical physics (Newton's, Maxwell's, Einstein's), which does not include the uncertainty principle, works quite well for macroscopic objects.

Now, concerning the electron, we can reduce the uncertainty in its velocity by using a photon of smaller energy so that its push to the electron is gentler. But a photon's energy is inversely proportional to its wavelength: the smaller the energy, the longer the wavelength (the "redder" the color is), a relationship that brings me to step two. Unfortunately, while a photon with a longer wavelength has less energy, which reduces the uncertainty in the velocity of the electron, it simultaneously increases the uncertainty in the position of the electron—the image of the electron gets fuzzier. Why?

Step Two: The Microscope

Because the determination of the position of the electron depends on the wavelength. This dependence, which is known as the resolving power of a microscope, regulates how clearly something can be seen—how well the scattered light can be focused and thus how accurately the electron can be located. The longer the employed wavelength, the fuzzier the image of the electron will be, and the greater the uncertainty in its position. What we see through the microscope is really a fuzzy flash created from the photon scattered by the electron. The electron, which is a point-particle, is somewhere within this flash, but where exactly is indeterminable. Its position cannot be known precisely. The flash may be focused into a region no smaller than the wavelength of light (the law of the resolving power states). Hence, the uncertainty in the position of the electron may be equal to or greater than the wavelength of light, but never smaller than it! So at best, the minimum uncertainty is equal to the wavelength: we cannot magnify (zoom in at) a region which is smaller than the wavelength of light we use to see something. Since the position cannot be known precisely, the electron has a range of possible locations it can occupy, just as it has a range of possible recoil velocities to move with.

Given that light of zero wavelength does not exist—that is, we cannot observe if the light source is turned off—the uncertainty in the position can never be zero: we cannot see with *absolute* precision where the electron is. Nonetheless, we can reduce the uncertainty in the position by using light of a smaller wavelength, though unfortunately this action simultaneously increases the uncertainty in the velocity—for as seen in step one, the smaller the wavelength, the greater both the light energy and the disturbance imparted on the electron (i.e., the greater the range of possible recoil velocities).

Position-Velocity Uncertainty

The wavelength of light used in an observation has conflicting effects; there is a trade-off in the determination of the position and velocity of a particle. The result is the position-velocity uncertainty principle: the more precise the position, the more uncertain the velocity, and vice versa.[7] Heisenberg proved mathematically that the product of the two uncertainties can never be less than a certain minimum positive number—which is roughly equal to Planck's constant, a fundamental constant of nature, divided by the mass of the particle.[8] Consequently, *absolutely* precise knowledge of either property is unattainable because if one of the uncertainties were zero, their product would also be zero, a result that would be in clear violation of the principle. In classical physics, on the other hand, of which the uncertainty principle is not part, these uncertainties could each be zero—thus, a particle's position and velocity could, at least in principle, be determined exactly—leading to what is known as classical determinism, which is the opposite of quantum indeterminism (that is, quantum probability), the consequence of the uncertainty principle.

Classical Determinism Versus Quantum Probability

In the macroscopic world of classical physics, by knowing the forces that act on an object as well as the object's exact position and velocity at some initial time, we can determine its exact position and velocity (its trajectory) for all past and future time. So its motion is precisely determinable: a path can be plotted, even watched live, point by point continuously from an initial instant to any future one. Because of this capability, classical physics is said to be deterministic. We can plot the precise orbit of a space shuttle, for example, just by knowing the forces acting on it and the initial conditions (its position and velocity at some initial instant), and we can watch it fly through space and time as predicted by our equations. It is therefore easy to predict a solar eclipse—when the earth, the new

---

[7] That is, it matters what you do first; measuring the position affects the velocity, measuring the velocity affects the position. Thus, the commutative law doesn't apply here: position × velocity ≠ velocity × position.
[8] Planck's constant is a very small number equal to 6.63×10^{-34} joules × seconds.

moon, and the sun will align—but absolutely impossible to measure or predict where an electron in an atom was is or will be. Why?

According to quantum theory, the subatomic world of particles is profoundly different than everyday experience; it cannot be described by classical physics. Inherent in the uncertainty principle, which limits the accuracy of a measurement, particle properties (such as position, velocity, momentum, and energy) cannot be assigned an exact value, neither initially nor at any time later. Thus, they must unavoidably be expressed only as probabilities, which then lead to quantum indeterminism. The wave functions, which are solutions to the so-called Schrödinger equation, can be used to calculate such quantum probabilities—for example, the probability of finding a particle at a certain location at a certain instant of time. A probability is a number that represents the tendency (the potentiality) of an event to take place, not its actual occurrence (the actuality). Hence, the best we can do is to theoretically predict only *probable* outcomes and experimentally measure the one outcome that actually occurs, although even then, our experimental knowledge is limited by the uncertainty principle. Consequently, a particle's path of motion can neither be predicted (nor plotted), nor can it be observed; it is indeterminable: establishing a definite, traceable, point-by-point orbit is an impossibility. In fact, the very notion of an orbit is inadmissible in quantum theory. Whence tiny particles come and whither they go can never be known. The determinism of classical physics is therefore replaced by the probability of quantum theory. And the consequences of this fact are staggering (affecting even human free will; see chapter 14). The true nature of nature, for example, is different ("disconnected") than the way it appears to be (continuous) through mere observation.

Observations Are Disconnected Events

During the act of observation, all we see through the microscope is a flash of light somewhere within which the particle exists. But where exactly it is within this flash at each instant, and what it does when we are observing it, whether at rest or in motion, are all indeterminable. Even worse, nature does not allow us to know what happens between consecutive observations. Consecutive observations have time and space gaps; flashes are seen one at a time and spatially separated. Hence, inherent in the uncertainty principle observations, *any* observations of both the microscopic and macroscopic world, are always disconnected events! Roughly speaking, it is as if we are observing nature by continuously blinking— our observations are dotted, intermittent, quantum!

This is a profound result and in direct contradiction with apparent reality according to which the changes in the daily phenomena are observed to occur continuously. The act of watching an arrow in flight, for example (an interesting thought experiment to be revisited in chapter 9) is really a series of disconnected observations, which to our imperfect eyes appear to occur continuously only because the time and space gaps between subsequent observations are undetectably short. The arrow's apparent continuity of motion is therefore an illusion. The shortness in these gaps is, incidentally, a consequence of the fact that in an observation, the disturbance introduced by a photon bounced off a macroscopic object such as car or an arrow is undetectably small because these objects have comparatively more mass (inertia) than the mass of microscopic objects such as electrons and protons. This is, recall, also the reason that macroscopic objects have (actually, appear to have) definite orbits while microscopic objects do not.

So observing anything, anything at all, can happen only discontinuously. It is, roughly speaking, like cinematography (motion pictures), where a series of separate drawings, each, say, of a ball at a different position, is flashed before us rapidly (with short time gaps). Now, (1) if the position of the ball is changed *gradually* in each subsequent drawing, that is, the distances (the space gaps) between each new position of the ball and the previous one are sufficiently short, then, when the drawings are flashed before us, the ball is observed to move continuously (thus with a definite orbit). But this continuity in observation is really an illusion of the deceptive senses that cannot notice the short gaps. Case (1) corresponds more to how we observe macroscopic objects. On the other hand, (2) if the space gaps are sufficiently long, then, when the drawings are flashed before us, the ball is observed to move discontinuously. Case (2) will correspond more to how we observe microscopic particles, but only after the following two modifications: first, do not think of the ball to be the actual particle; rather, it roughly corresponds to the flash of light (the wavelength) somewhere within which the observed particle exists; and second, as we will argue in "Nature as a Process" later, even if we do observe a similar *type* particle (say, an electron) at consecutive observations, it is indeterminable if it is the same *particle* (electron), even when the time and space gaps between observations are short. These two modifications, which capture more accurately how we observe microscopic particles, make it impossible to plot a definite path of motion for any microscopic particle.

Now, the reason that the phenomena are observed to occur discontinuously might be that the very phenomena *themselves* occur discontinuously (even when we are not observing); they may not just be *observed* to occur discontinuously. In any case, the discontinuity in observations has astounding consequences: in the section "Nature as a Process," we will use it to question the very identity of a particle, and in chapter 9 we will use it to question the reality of motion *itself*. But to understand such consequences we must first change the topic.

Change

The Heraclitean doctrine that everything is constantly changing and nothing is ever the same has three implications. First, there is a change; second, the change is constant; and third, because nothing is ever the same, the constant change is unidirectional. (That's why we learn calculus, the mathematics of change.) Modern physics agrees with all three: first, change occurs in two different ways: (1) through motion and (2) through the transformations of matter and energy; second, the uncertainty principle of quantum theory and the theory of general relativity affirm that change is constant,[9] and in addition, as seen just earlier, quantum theory ascertains it is also discontinuous; and third, the second law of thermodynamics discusses how change is unidirectional—the universe becomes increasingly disordered.

(1) Motion Causes Change

Change caused by motion is discussed in the following three cases.

A. The motion of the particles of matter causes their rearrangement in an object and consequently causes change in its various qualities (e.g., density and temperature). For example, atoms are more compressed in denser objects and jiggle faster in hotter objects.

B. Space itself is changing because matter distorts it, a phenomenon that can be understood when we describe how gravity works within the context of the theory of general relativity.

Gravity, in general relativity, is explained by giving space geometrical properties, namely, by regarding it as a flexible medium distorted by matter—like a trampoline surface that is stretched and warped by a bowling ball resting or moving on it. In the case of the earth and the sun, for example, the distortion of space caused by one body influences the motion and is felt as gravity by the other.

In a simplified analogy, the flexible trampoline fabric (which plays the role of space) is curved when a bowling ball (which plays the role of the sun) rests on it. The geometry (shape) of the fabric depends on (1) the mass of the bowling ball and (2) the distance from it: (1) the more the mass, the more curved the fabric (space) becomes; (2) the closer to the bowling ball, the greater is the curvature of the fabric. The distorted fabric in turn influences the motion of a small marble (which plays the role of the earth) rolling on it. Depending on how we start the marble moving (i.e., with what initial speed and direction, and from what location), it will move on the distorted fabric by following a particular path (circle,

---

[9] Although for the "block-universe" interpretation of relativity, change appears to be an illusion (see chapter 8).

ellipse, parabola, spiral, toward the bowling ball, etc.), and thus will *appear* to be attracted by the bowling ball. For example, a marble released from rest moves on the distorted fabric caused by the bowling ball and plunges onto it, like an apple falls from its tree onto the ground by moving through the distorted space caused by the earth. Newton thought the apple is attracted by the earth—although *how, really*, the apple knows that the earth is below in order to fall, he didn't know. Einstein figured it out because he reimagined gravity: just as the marble is pushed by the curved fabric of the trampoline and *appears* to be attracted by the bowling ball, the apple is pushed by the curved fabric of space-time—by the curvature, the geometry, of space-time—and *appears* to be attracted by the earth. Gravity is the geometry of space-time. In the trampoline analogy the distorted fabric *is* gravity; and the greater the bowling ball mass, or the smaller the distance from it, the stronger gravity (the distorted space-time) becomes.

In the earth-sun case, the sun distorts the spatial fabric around it (and time, too—it passes more slowly as one gets closer to the sun, or any other object in general). Traveling at the speed of light, this distortion reaches and affects the motion of the earth—analogously, a water disturbance, a water wave, travels in the sea but with a much smaller speed. In turn, the earth traverses the distorted space as if space pushes the earth through it. Of course, the earth distorts the space around itself, too (although its smaller mass produces a much smaller distortion than that of the sun); and so do the moon, planets, stars, and galaxies. Gravity is really the twists, curves, ripples, bumps, depressions, and in general all these distortions (the changing geometry) of space-time. And each body's motion is actually a response to the space distortions from all other bodies around it. Because these bodies are in constant motion in the universe, the pattern (the geometry) of space distortions that they create is in a state of constant flux—and so the motion of matter causes change in the geometry of space. In our analogy, as it rolls, the bowling ball transfers the warping of the trampoline surface to different locations. Because both space and time are distorted by matter, space-time in general relativity becomes a four-dimensional malleable (distortable) continuum. In turn, these space-time distortions, which we usually call gravity, influence the motion of matter.

C. In addition to its constant warping, space as a whole is also expanding according to the big bang cosmology, thereby carrying all the galaxies with it and causing them to move away from each other. Here again, motion, which in this case is a consequence of the expansion of space, produces change. Known as the expansion of the universe, this was first predicted theoretically by the solutions of the equations of general relativity, shortly after their publication in 1916. It was later confirmed experimentally by astronomer Edwin Hubble (1889–1953) in 1929 when he observed a redshift in the light emitted by the distant galaxies. The redshift is a measure of the relative velocity between a galaxy and the earth.

Specifically, it means that distant galaxies are rapidly receding from us. The greater the distance, Hubble discovered, the faster the recession speed, a result known as Hubble's law. This law is included in the big bang model.

Galaxies are not moving out into preexisting space, a common misinterpretation of the phenomenon of expansion, but they are moving away relative to each other, and what carries them is space itself as it is expanding (stretching); the result is that the size of the universe is increasing with time. Furthermore, the recession of galaxies does not make the earth the center of the universe or in any way a more special place than any other. Quite the opposite, because the universe is isotropic, the expansion would look the same from any location in the universe. In a classic analogy, imagine how any dot (i.e., galaxy) on an inflating balloon is seen receding from the perspective of any other dot as the expanding membrane (i.e., space) itself carries all dots with it. Since galaxies are observed receding from each other and the universe to be expanding, in the past they must have been closer to each other, and the universe must have been much smaller—imagine our balloon to be deflating. In the extreme case, the whole universe (all the galaxies, all matter, energy, and space, even time) is imagined to have been a mere point, its explosion of which is the premise of the big bang model. Whether such a point-like universe actually existed is still only a hypothesis, and why it exploded is still puzzling. Nonetheless, we do know that the universe must have been very, very small (if not point-like) when it exploded, as described by the big bang theory, and that it has been constantly expanding (stretching), thus changing ever since.

(2) Transformations Cause Change

Transformations of matter and energy also cause change in the various qualities of objects (or the universe in general) via two types of processes: first, when the various particles of energy convert into particles of matter, back and forth, materializing and dematerializing; and second, when one type of material particle converts into another. An example of the former is the materialization of the energy of invisible gamma rays into an electron-positron pair, or the dematerialization of such pair into the energy of gamma rays; and an example of the latter is the conversion of two protons into a proton-neutron nucleus (a heavy form of hydrogen called deuterium), a positron, and an elusive neutrino (all common reactions in the stars energizing them with their light). Of course, more everyday-type transformations of matter and energy, such as from solid to a liquid (as the melting of ice) or liquid to a gas (as the evaporation of water), also cause change.

But things are not merely changing; they are *constantly* changing, a conclusion required by the uncertainty principle.

Change Is Constant

To avoid violating the uncertainty principle, motion in nature must be perpetual. If a particle could sit still, it would mean that its velocity would be exactly zero and so then would be the uncertainty in its velocity. Consequently, the product of the position and velocity uncertainties would also be zero, a result in violation of the Heisenberg uncertainty principle. The principle holds only if motion is perpetual. A particle cannot sit still, ever. This result is also supported by the third law of thermodynamics, which states that the absolute zero—the lowest temperature for which every particle in a substance would have been motionless—is unattainable (temperature is a measure of particles' average energy of motion). Now, since the motion of particles is perpetual, so is change. Incidentally, since motion is constant, then motion is also involved even when change is caused by the transformations between the particles of matter and the particles of energy.

Support of the "constant change" view comes also from the expansion of the universe that has been happening ever since the big bang. What's more, the space-time continuum fluctuates constantly at microscopic scales, like a turbulent sea—the distortions of space are changing violently—a phenomenon resulting from efforts to reconcile the theory of general relativity with the uncertainty principle of quantum theory.

Change Is Unidirectional

But change is not merely constant; it is also unidirectional, meaning nothing is ever the same; "you could not step twice into the same river."[10] Heraclitus parallels the constant unidirectional change in nature to the ever-changing waters in a river. If the state of a river at one moment were ever the same as the state at another moment, it would have been possible for one to step twice into the same river. Since one cannot do that, nothing ever is the same, and so change is not only constant; it is also unidirectional. In fact, one "could not step twice into the same river" not only because a river's waters are ever-changing, but also because one's own body is also ever-changing. *Everything* is constantly changing, and *nothing* is ever the same. Parenthetically, we think

---

[10]   Plato, *Cratylus* 402a8–10, trans. Graham, *Texts of Early Greek Philosophy*, 159 (text 63).

we are the same individual throughout life, but I have always wondered, what happened to that 10-year-old boy that I once was? I'm obviously (physically) not the same, but am I still the same individual (being)? Einstein, we'll see in the next chapter, claims that every moment of my existence (in some sense) still exists, and such a surreal claim has the support of the bold logic of Parmenides!

Now, according to the second law of thermodynamics, net entropy—the degree of disorder (randomness) in the universe—is always increasing. Think of the universe as a jigsaw puzzle. There are more disorderly arrangements than ordered ones (and only one of the perfect picture). Had I dropped the pieces on the floor, the likelihood to form a disorderly pattern is much higher than that of the perfect picture. This likelihood becomes ever more accurate with ever more jigsaw pieces. The universe, like the puzzle, has vastly more disorderly configurations than ordered ones, so naturally with time the universe's entropy increases. Thus, nature is in a state of becoming, but it is a disorderly becoming.

Indeed, then, everything is constantly changing, and nothing is ever the same! Heraclitus's doctrine of change includes everything, even the seemingly unchanging, such as a rock in its apparent state of rest or even a human body. In fact, even the gradual biological evolution by descent and variation—that the more complex life forms do not arise spontaneously but evolve from simpler ones through modifications—is a principle to be expected as a consequence of the Heraclitean theory of constant change.

Nature as a Process

Heraclitean View

A profound consequence of the Heraclitean theory of universal constant change is the view of nature as a process made up of events. For the notion of "a thing" is inconsistent with a theory of constant change. To be able to be spoken of and defined, the thing must remain absolutely the same for at least a period of time; it must have some permanence and must be identifiable. But the notions of sameness and changelessness are contradictory to a theory of *constant* change. Consequently, it is more appropriate to consider a thing as an event (something happening somewhere at some instant of time) and not as something permanent. Thus, what changes is not something material or initially permanent; what changes are the events. Groupings of events constitute processes, which in turn make nature the ultimate process.

Quantum View

This notion is supported by quantum theory. We will argue that microscopic particles are better understood to be events rather than permanent things.

We learned earlier that because of the uncertainty principle observations are disconnected events. Now, without continuity in observation, without the ability to keep a particle under continuous observation (even for the smallest time duration), how can we establish its identity or permanence? With time (and space) gaps between observations during (and within) which we cannot see what a particle is doing, how can we be sure whether, say, an electron observed at location A has moved there from location B, or whether it is really one and the same electron as that observed at location B, regardless of their proximity? We cannot! Since observations are disconnected events, consecutive observations of identical particles—such as electrons, all of which have the same intrinsic properties, for example, charge, spin, rest mass—might in fact be observations of two different particles belonging in the same family (e.g., two different electrons), and not observations of one and the same particle (e.g., the same electron) that has endured for a certain period of time. It is therefore impossible to ever determine whether the observations of two identical particles could actually be observations of one and the same particle, and consequently whether a particle endures for a period of time.

So without the ability to keep a particle under continuous observation, it is impossible to establish experimentally its identity or permanence. Because of this, the notion that a particle is an identifiable individual and a permanent *thing* breaks down (or, it is an ambiguous notion, to say the least). The alternative is to consider a particle to be an event.

General Relativity View

Particles, in the view of general relativity, can endure up until they convert to energy. Until then they are identifiable permanent entities because general relativity has not yet been reconciled with the uncertainty principle. Still, particles are events in general relativity, too. This is so because matter is intricately connected with the fabric of space-time (they are continuously affecting each other). So as time is constantly changing, so are, in general, the properties of space and matter. Hence, a point in the continuum of space-time is regarded as an event. And so a particle occupying a space location at an instant of time is treated also as an event. Two events are separated by their space-time interval, which involves a spatial distance and a time interval.

In conclusion, matter, energy, space, and time are all intimately linked, interacting with one another constantly, causing changing events and processes. Nature is the perfect process and therefore is in a state of becoming; nothing ever is. But what causes change in the theory of Heraclitus, and what causes change in modern physics?

Fire and Energy

A permanent primary substance of matter is contradictory to a theory of constant change. The only element of permanence in such a theory is change itself. What really causes change then? For Heraclitus it was the "everlasting fire,"[11] and for modern physics the eternal energy. The strife of the opposites or the interactions of matter are fueled by fire and energy, respectively. Matter, the events as has been argued, is the result of the transformations of the fire or of the energy. Fire and energy also represent particular processes. They cause cooling and condensing or heating and rarefying or forming and dissolving. "The transformations of fire [energy] are, first of all sea [liquids]; and half of the sea is earth [solids], half whirlwind [gases]."[12] And the transformations of fire, as it is with energy, occur with measure—by obeying conservation laws—since in a metaphor Heraclitus argued, "All things [matter] are an exchange [is a transformation] for fire [of energy] and fire [and energy is a transformation] for all things [of matter], as goods [as one type of matter] for gold [transforms into another type of matter or energy] and gold for goods [back and forth]."[13] In another statement he writes, "This world, which is the same for all, none of the gods nor of the humans has created, but it was ever and is and will be, an everlasting fire [energy], flaring up with measure [conservation] and going out with measure [transforming back and forth between its various forms by obeying the law of conservation of energy]."[14] Change, which in each respective theory is caused by fire or energy, is guaranteed to always occur since fire and energy are conserved while they are also changeable. From all these qualities fire and energy seem to qualify as the primary substance of the universe in each theory. As argued by Popper, if the world is our home, then for Heraclitus fire is not *in* the house, "the

---

[11] Clement, *Miscellanies* 5.103.6, trans. Demetris Nicolaides. Or see Graham, *Texts of Early Greek Philosophy*, 155 (text 47).
[12] Ibid., 5.104.3–5, trans. John Burnet, *Early Greek Philosophy* (London: A & C Black, 1920), chap. 3.
[13] Plutarch, *On the E at Delphi* 338d-e, trans. Graham, *Texts of Early Greek Philosophy*, 157 (text 55).
[14] Clement, *Miscellanies* 5.103.6. Or see Graham, *Texts of Early Greek Philosophy*, 155 (text 47).

house [the *house*] is on fire," a "somewhat more urgent message."[15] Equivalently, we can argue, the world is energy.

Fascinated by the similarities between the Heraclitean fire and energy, Heisenberg wrote: "Modern physics is in some way extremely near to the doctrines of Heraclitus. If we replace the word 'fire' by the word 'energy' we can almost repeat his statements word for word from our modern point of view."[16] And also, "Energy is in fact that which moves; it may be called the primary cause of all change, and energy can be transformed into matter or heat or light. The strife between opposites in the philosophy of Heraclitus can be found in the strife between two different forms of energy."[17]

Organization

But while everything is constantly changing, and defining a permanent fundamental particle of matter is impossible, something must endure, at least for some time. For, in all this constant change, still, definable "things" such as rivers exist and so do we, and rivers are distinct from us, and one river (or one person) distinct from another. So both we and the rivers are constantly changing, but simultaneously both we and the rivers are recognizable. "We step and do not step into the same river; we are and are not,"[18] said Heraclitus. How can this be?

In a constantly changing nature, which is best regarded in terms of events, each event is different. But *collections* of events can endure, by creating a certain macroscopic average (an emergent reality), which, for a period of time, is a recognizable "organization";[19] the identifiable plethora that we call things may then be viewed as different organizations.

From the earlier Heraclitean quote, in the part "we step . . . into the same river; we are . . . ," Heraclitus treats the "we" and "river" as two identifiable organizations, thus as two different and permanent things—at least for some period of time and in some region of space. Whereas in the part ". . . and do not step . . . and are not," he seems to imply that while the "we" and the "river" appear

---

[15] Karl R. Popper, *Conjectures and Refutations: The Growth of Scientific Knowledge* (London: Routledge, 1989), 147.

[16] Werner Heisenberg, *Physics and Philosophy: The Revolution in Modern Science* (New York: Harper Torchbooks, 1962), 37.

[17] Ibid., 45.

[18] Heraclitus, *Homeric Questions* 24, trans. Demetris Nicolaides. Or see Graham, *Texts of Early Greek Philosophy*, 159 (text 65).

[19] Schrödinger, *Nature and the Greeks*, 123–125.

as permanent things, these are so only on the average. In reality, neither "we" nor the "river" is ever the same.

Heraclitus realized that while a thing (an organization) can on average be identifiable for some period of time, strictly speaking the thing is uniquely different at each instant of time. And so while there may be a collection of events that can produce an identifiable organization called river, which on average persists unchanged for a certain time duration and in some region, this river is never ever really exactly the same. Logos (the cause of change, the law of nature), in the view of Heraclitus, is the only thing truly eternal. Although through a truly strict interpretation of a continuously changing nature without anything ever the same, even Logos should be changing. The modern physics equivalent of this is the hypothesis that the fundamental constants of nature, numbers that describe the various laws we have and are the reason the universe is what it is (such as the speed of light, Planck's constant, or the gravitational constant), might after all be functions of space and time. If this is discovered to be true, then the order and organization of tomorrow's nature (especially in the long run) will be so unknowingly different from today's, a real intellectual treat for those inquiring minds who like the constant search and discovery and the journey more than the destination, for in such a case, the mysteries concerning the nature of nature will be forever changing, and so will our very knowledge about them.

Conclusion

Heraclitus declares the being (that which exists, nature) but identifies it with becoming. All follows from that: everything is constantly changing, material sameness is impossible, there is a plethora of different events that make nature a process, and described by warring opposites that nonetheless obey Logos. But Parmenides declares just the *Being*; only what is, is, and what is not, is not. All "follows" from that: change, he argues, is logically impossible and so what is, is one and unchangeable! This dazzling and absolute monism is in daring disagreement with sense perception but curiously it has found a well-known genius, Einstein, as a supporter.

The Heraclitean and Parmenidean worldviews are therefore antinomies (contradictions), for starting from a being the two philosophers developed a unique series of logical arguments and arrived at opposite results: for the Heraclitean, being is becoming, but for the Parmenidean, Being just is. It is Heraclitean change and plurality versus Parmenidean constancy and oneness. But it is a controversial constancy and oneness, for Being's exact nature is uncertain.

8

The Unchanging Universe

Introduction

Philosophy was seriously shocked by the logic of Parmenides of Elea (ca. 515–ca. 445 BCE). Being the first philosopher of ontology, Parmenides questioned the nature of existence itself and created his monistic philosophy by contemplating the most fundamental of questions: How can something exist? And what are the properties of that which does exist? And through purely rational arguments he marvelously reasoned out an answer that overturned completely the common perception of the world around us! In particular he asked, How could there be something instead of nothing? What does it mean to say that something exists? Can existence (nature) come to be from nothingness? Is there such a thing as nothingness? Has nature been caused by a primary cause—that is, by an absolutely first cause that permits no cause (no explanation) of its own? Does nature have an ultimate purpose that permits no purpose of its own? What is the nature of nature? Remnants from his profoundly abstract thought are present in modern cosmological models describing one indivisible and whole universe, unborn, eternal, imperishable, even unchanging.

Ironclad Logic

First he argued that we can think only about that which exists, the *Being*, "for the same is the thinking and the Being."[1] (Descartes's famous "cogito ergo sum," I think, therefore I am, is Parmenides's quote in disguise.) On the contrary, he thought, we can neither speak about nor think about something that does not exist, *Not-Being*. For if we could, it would mean that Not-Being had properties (those mentioned speaking or thinking about it). But true nothingness is property-less. Therefore, the notion of nothingness (Not-Being) is impossible! This is in fact the critical premise of Parmenides's theory. And to understand his arguments we must always remember that for him what does not exist, does not exist, neither now nor before or after, neither here nor there; that is, we cannot

---

[1] Clement, *Miscellanies* 6.23, trans. Erwin Schrödinger, *Nature and the Greeks and Science and Humanism* (Cambridge: Cambridge University Press, 1996), 27.

assume what does not exist now (or here), could exist later (or there), or could have existed before somewhere. No! Only what is, is—only Being exists. What is not, is not—Not-Being does not exist.

With this premise in mind, he proceeded to figure out if change is logically possible. Change, he maintained, requires that the notion of nothingness exists. But since such a notion is an impossibility, so then is change; Being is unchangeable—for if it could change, it would change into something that Being is not already, into something new that does not yet exist, thus into Not-Being, but this is an impossibility, for Not-Being does not exist ever anywhere. Analogously, if it could change, it would cease to be what it once was, thus what once existed (Being) would no longer exist; it would become Not-Being, but this is again an impossibility for Not-Being does not exist. In other words, that which exists (Being) cannot change because change requires that the notion of nothingness (Not-Being) exist. Because only then could Being have been it (Not-Being) and could have again become it. Simply put, we can say that change is impossible because it requires that something is either created from nothing or destroyed into nothing, but since the notion of nothingness does not exist, change does not exist either.

Being and the Block Universe

Does the universe change or not according to modern physics? In his deterministic conviction of nature, and emboldened by his special and general theories of relativity, Einstein considers the universe as a four-dimensional "block" (a spacetime continuum like a loaf of bread[2]) which, remarkably, contains *all* moments of time (of past, present, and future) *always*, and where change is an illusion. He said, "For we convinced physicists, the distinction between past, present, and future is only an illusion, however persistent."[3]

Specifically, the notion that the present, what exists, becomes the past, what does not exist (that, in Parmenidean logic, what exists, the present, Being, can become the past, what does not exist, Not-Being) and that the future, which does not yet exist, becomes the present (that, in Parmenidean logic, what does not yet exist, the future, Not-Being, can become the present, what exists, Being) is true in the Newtonian and the quantum-mechanical universe, but it is false in the Parmenidean theory (for Being cannot become Not-Being and vice versa). It is

[2] Although the bread is three-dimensional (with two space and one time dimensions), the real universe is four-dimensional (with three space and one time).
[3] Albert Einstein quoted in Brian Greene, *The Fabric of the Cosmos: Space, Time, and the Texture of Reality* (New York: Vintage, 2005), 139. Also quoted in Carlo Rovelli, *Seven Brief Lessons on Physics* (New York: RiverHead Books, 2016), 60 (Kindle ed.).

also false in Einstein's block universe, because in relativity not only all locations of space exist always, but so do all moments of time, for space and time have no separate existence; rather, together they weave the very fabric of the continuum of space-time (the bread). Now, since *all* moments of time exist in the continuum *always*, they neither come to be (as if Not-Being became Being) nor cease to be (as if Being became Not-Being)! Parmenides might be smiling, I mean *really*. But how?

In a concrete example, according to relativity (for which time is *relative*), I'm still a baby relatively to a woman moving away from me[4]—namely, my past is part of her present;[5] and my present will be part of her future. Yet, relatively to a man, moving toward me, I am the older person I will grow up to be—namely, my future is part of his present; and my present is part of his past. Generally, every event (space-time point) of my life, or even more generally, *every event*, period (like a smiling Parmenides), of past, present, and future, is always part of the space-time continuum—it exists as Being! Or *to say the least*, every event is part of someone's present and thus it exists—it is Being!

The fact that *all* of space and *all* of time exist *always* justifies the terminology *space-time continuum*—or block, or loaf-of-bread universe for which *all* breadcrumbs (the "events") are *always* in the bread (the space-time continuum). Since all events exist always, relativity then is in accordance with one of the Parmenidean inferences, that change is an illusion: things, or events, neither come to be (as if Not-Being became Being) nor cease to be (as if Being became Not-Being). What is, is, and what is not, is not. In the block universe, the present doesn't change into the past, and the future doesn't change into the present; to the contrary, in the block universe, the past is not gone, it is present; and the future, like the present, is, well, present, too.[6] Time, in the continuum, doesn't flow, for all of it, is always there. "I [Popper] tried to persuade him [Einstein] to give up his determinism, which amounted to the view that the world was a four-dimensional Parmenidean block universe in which change was a human illusion, or very nearly so. (He agreed that this has been his view, and while discussing it I called him 'Parmenides'.)"[7]

---

4   With the appropriate speed and distance—numbers calculated by the equations of relativity.
5   That is, my baby-self is a simultaneous event relatively to her, not to me. The expression, "you live in the past," suddenly makes more sense.
6   If change is truly an illusion, how is it created giving us a personal view of the universe in which we remember/know only the past (and only part of it) but not the future (an experience that has been called psychological time)? It's an open question. To the contrary, the moon's apparent small size, also an illusion, can be explained: it's due to its distance.
7   Karl R. Popper, *Unended Quest: An Intellectual Autobiography* (London: Routledge Classics, 2002), 148; or see: https://books.google.com/books?id=NyCEnehPMd8C&pg=PA148&lpg=PA148&dq=popper+einstein+parmenides&q=&hl=en#v=onepage&q=popper%20einstein%20parmenides&f=false (accessed January 9, 2017).

Interestingly, diversity does exist in the block-universe interpretation since each space-time point is a unique event. But there is no change since all such diverse events exist unchangeable. The block universe is like a painting: all its points exist unchangeable, but they are also uniquely different, and together all form a beautiful diverse union, the canvas. Parenthetically, one interpretation of Parmenides's universe might be so too, unchangeable but diversified.

Uncaused

Being is also unborn (it is uncaused; that is, it has not been caused by anything and thus has no beginning) and it is imperishable (it has no end, no ultimate purpose). It just is! It could not be born, Parmenides thought, for if it could, it would be born from either (a) Not-Being, but this is an impossibility, for Not-Being does not exist; or (b) Being, which is also an impossibility, for something cannot be born if it already exists; that is, something cannot be born from itself. Analogously, Being cannot perish; for nothingness, which Being must become in order to perish, does not exist. Hence, what is (Being) just is; it neither comes to be from nothing nor perishes into nothing. This remarkable Parmenidean thesis was embraced by the pluralists Empedocles, Anaxagoras, and Democritus, and as we will see, it was applied in their own theories. Well, then, since Being is unborn, unchangeable, and imperishable, there is neither becoming nor passing away—nature just is!

Timeless and Omnipresent

Being is also always everywhere: there is neither a place nor a time where or when what is, is not already complete (e.g., of the same amount, appearance, and generally of the same properties). For if somewhere or sometime Being were less than complete, if it lacked something, it would mean that somewhere or sometime that something that Being would lack would not exist; it would be Not-Being, but since Not-Being does not exist, there is never any expectation for Being to be it. So Being is always complete everywhere. Hence, diversity and plurality are illusions of the senses.[8] It is also motionless—for being always everywhere, there

---

[8] In my opinion, Being's completeness property may also be interpreted as the "painting" analogy of the block universe (see end of section "Being and the Block Universe"), where each space-time point is "complete" and unchangeable but it is also different than all other space-time points, creating an unchangeable but diversified universe.

is never anywhere to go where it is not already. Similar-type arguments lead to the various properties of Being.

Oneness

The nature of nature (of Being) is of the purest oneness: there is only one thing, Being. It is an indivisible eternal whole, unborn, unchangeable, imperishable, continuous, indestructible, finite, and uniform (always the same everywhere). This is a dazzling but provocative oneness, for it is (or so it seems, anyway) logically sound, yet it is also daringly in stark contradiction to apparent reality. And what its exact meaning really is depends on how these properties of Being are regarded, literally or metaphorically, of material or immaterial nature. Being is that which is. All other characteristics beyond that are quite uncertain and debatable, for what *is* (what exists) is *really* the question for Parmenides. But irrespective of its nature, Being captures a highly valued place in ancient natural philosophy and modern physics alike—namely, oneness!

Modern physics embraces a kind of monism and wholeness, too, for it tries to ultimately unify all four forces (gravity, electromagnetism, the nuclear strong, and the nuclear weak) and all particles under a single overarching principle in which there will be only one unified force or, equivalently, one type of fundamental particle, suggesting a subtle interconnection and oneness in all apparent plurality. Hence, should the properties of Being be interpreted metaphorically, Being might be a metaphor of the one, unchangeable, universal, eternal, objective truth of nature (a unified force or grand idea of a theory of everything).

On the other hand, should the properties of Being be interpreted literally, then nature is one, uninterrupted, indestructible, indivisible, eternal, and material whole; it's a kind of full and solid block of matter without parts (uniform). Such a type of full nature implies that there is no void (empty space). The void is Not-Being for Parmenides, true nothingness—it does not exist. But interestingly without the void it is difficult, if not impossible, to accept that motion and change are real. For the easiest way to understand the occurrence of motion and thus change, too, is to imagine the existence of empty space within which things could move. Assuming there is no empty space, motion is an illusion and so is change. The atomists Leucippus and Democritus found this conclusion utterly absurd but inspiring as well. For to create their atomic theory and explain motion and change rationally, they had to employ both: Parmenides's Being—its material (solid-block-like) nature in particular—and his Not-Being.

Atom and Being, Void and Not-Being

So by giving Parmenides's ideas a straightforward, literal, and material meaning, the antithetical notions of his Being (existence) and Not-Being (nonexistence) evolved in the minds of the atomists into the antithetical notions of "the full"[9] (the atom) and "the empty"[10] (the void, empty space), respectively, and became the essence of their atomic theory. Incidentally, the intellectual continuity in our efforts to know nature is unquestionable in this case. Now, there were many atoms (Beings) with key properties of Being (i.e., whole, indivisible, indestructible, solid, with no parts) and lots of empty space (Not-Beings) within which atoms can move. Interestingly, although the atomists could not counter Parmenides's arguments against the existence of Not-Being, they still identified it with a certain kind of nothingness that *existed*, the empty space. But empty space's perception has been controversial ever since its conception.

Nothing Comes from Nothing

For Parmenides, there is no empty space, for empty space is nothingness, Not-Being, and Not-Being does not exist. How can something, which is nothing, really exist? Parmenides thought. How can something be defined and assigned properties when it is supposed to be property-less? It cannot, he argued. We are unable to even think of nothingness, he reasoned. Nothingness is a meaningless concept, for if nothingness existed, it would not really be nothingness; it would be something-ness. If we could refer to something and give it properties, that something could not be nothing; it would be something real and would exist.

Parmenides wanted to understand change, motion, and the empty space via purely logical arguments. For the empty space especially, he thought there was no good logical argument in support of its existence. Whether the empty space is a true nothing or not is a notion to be revisited in chapter 12. Nonetheless, Parmenides thought that empty space was a true nothing, and as seen earlier, he also argued that *nothing* comes from nothing—Being neither comes to be from nothing (it is unborn) nor passes away into nothing (it is imperishable).

This principle has in fact a counterpart in modern physics in the notion of energy, which includes mass, since for relativity mass and energy are basically the same, as implied by $E = mc^2$. Like Being, energy can neither be created (it does

---

[9] Aristotle, *Metaphysics* 985b4–20, trans. Daniel W. Graham, *The Texts of Early Greek Philosophy: The Complete Fragments and Selected Testimonies of the Major Presocratics* (Cambridge: Cambridge University Press, 2010), 525 (text 10).
[10] Ibid.

not come to be from nothing) nor destroyed (it does not pass away into nothing). It just is and its total amount is unchangeable and enduring. Particles of matter and antimatter are constantly being created and annihilated, for example, but not out of nothing and into nothing. To occur, these processes require something to already exist—namely, energy. They are created from energy and annihilated into energy. There is no mechanism in modern physics that violates the basic Parmenidean idea that something can neither be created from nothing nor pass away into nothing. All interactions require something, energy (or matter, since they are equivalent), but also space and time. In fact, while space contracts and time dilates in Einstein's relativity, the combined *space-time* interval is invariant (conserved), too (like the combined matter-energy). Indeed, nonexistence is an impossibility even in modern physics, and the uncertainty principles of quantum theory (see next paragraph) may be considered as additional statements (in addition to those of relativity) in support of that. Why use these principles? These principles are relationships between space, time, matter, and energy, concepts that constitute the essence of nature (of something-ness). And if we hope to "prove" that the notion of nothingness is an impossibility—that nothingness is not derivable from something-ness—well, we had better begin from an analysis of principles that describe the essence of something-ness.

So to argue for this (that nonexistence/nothingness is an impossibility), we first recall the position-velocity uncertainty principle: the product of the uncertainties in the position and in the velocity of a particle must be greater than Planck's constant divided by the particle mass—that is, such product is greater than zero. Analogously there exists the time-energy uncertainty principle. It states that the product of the fluctuations in the energy of a particle and the time interval that the particle endures must be greater than Planck's constant—again such product is greater than zero. For these uncertainty principles to hold spatial distances, time intervals, velocities, and energies are forbidden from ever being absolutely zero—that is, their nonexistence is forbidden. For example, the smaller the confining space of a particle is (or the briefer the time interval the particle endures in such confinement), the more frantic its motion and energy are. But neither the confining space nor the time interval can ever be exactly zero, for if they were, the uncertainty in position and the uncertainty in time would have been zero, too, and consequently both uncertainty principles would have been in violation—the product of the uncertainties in position and velocity, and in time and energy, would have been zero, too (instead of greater than zero). Similarly, both uncertainty principles would have been in violation if a particle had zero velocity or energy. The principles hold only if spatial distances, time intervals, velocities, and energies are nonzero; they must exist; they cannot be nothing! (In a sense, such a result is expected because our physics relationships, equations or inequalities, are in the first place conceived to describe *something*, not nothing;

the notion of nothingness is indescribable.) Hence, as per Parmenides's reasoning and as per the uncertainty principles, nothingness is not only not allowed to exist—for *nothing* comes from it (e.g., the uncertainty relationships are violated and thus cannot be used to account for what exists)—but, equally profoundly, existence is required, that is, spatial distances, time intervals, velocities, and energies must be nonzero (for only then do the uncertainty relationships, which describe something-ness, hold).

In fact, one of the fundamental tenets of quantum theory is that information cannot be lost from the universe (recall the section "Black Holes: Challenges in the Quest for Sameness" in chapter 3). In Parmenidean terms this means that what is (Being, information already present) cannot become Not-Being (information cannot be lost).

Quantum theory (the essence of which is the uncertainty principles) is then in accordance with Parmenidean philosophy, "for the same is the thinking and the Being": for we can think only about that which exists; in other words, the uncertainty principles describe only something-ness and forbid nothingness. With this in mind Parmenides's main question, How can something exist? may now be answered: within the context of modern physics, something (Being) must exist because nonexistence (Not-Being) is impossible.[11] Although still, science cannot answer why *this* universe with *these* laws, why not some other type? Anyhow, what is the nature of that which exists?

An Indivisible Whole

Relativity

The view of Being as an indivisible whole is supported by Einstein's theory of general relativity: for *everything that there is*, space, time, matter, and energy are no longer independent of each other (that is, they are not absolute), as was the case with Newtonian physics, but are intimately interwoven, affecting one another constantly. "Time and space and gravitation have no separate existence from matter."[12] Space-time is a malleable continuum distorted by matter.

Yes, it is true that for the sake of practical calculations in physics we often isolate, in our mind, a phenomenon of interest by assuming that it is disconnected from the rest of nature (disconnected from the whole). For example, we study the

---

[11] It's worth emphasizing once more that this answer is expected, for as mentioned, the uncertainty principles, which helped us to arrive to this answer, are in the first place conceived to describe *something*, not nothing; the notion of nothingness is indescribable.

[12] Einstein quoted in Joanne Baker, *50 Physics Ideas You Really Need to Know* (London: Quercus, 2007), 165.

gravitational interaction between the sun and the earth by neglecting the gravitational effects of the rest of the heavenly bodies. But as in the philosophy of Parmenides modern physics is about oneness, not isolation. And in reality all things in nature are part of the whole and are entangled far more intricately than the theory of relativity alone could discover.

Quantum Entanglement

One of the most fascinating consequences of quantum theory is the phenomenon of quantum entanglement. According to it, there are no perfectly isolated particles (or systems). The notion of an individual particle disconnected from the rest of the universe is inaccurate. Rather, all particles in the universe are part of a unified whole. They are in constant and *instant* interaction, affecting and determining the behavior of each other regardless of how far apart they are. Quantum theory suggests that everything that happens in the universe influences instantly everything else. In this sense the universe is indeed a Parmenidean indivisible whole. To explain this concept further, we use the following thought experiment.

Suppose, for simplicity, that a mother particle could initially be at rest and with zero spin, and that later it decays into two daughter particles, A and B. To conserve momentum (linear and rotational), the daughter particles must take off away from each other as well as spin in opposite directions. In 1935, Einstein, with Boris Podolsky (1896–1966) and Nathan Rosen (1909–1995), argued through this thought experiment (which is known as the EPR, from the initials of their last names), that the daughter particles must have a fixed spin *since the moment of their creation*. To conserve rotational momentum, one must spin clockwise and the other counterclockwise. Which particle spins in what direction is determined with a measurement. So if Alice measures that particle A spins clockwise, she is also certain that particle B must spin counterclockwise, as it is so confirmed when Bob measures it. Einstein's view is really the deterministic view of classical physics: that a particle has a fixed property *even before we measure it.*

But according to quantum theory, the spins of the particles A and B become fixed *only when an observation (a measurement) is made.* Until then, not only do we not know how the particles spin, but even worse and unlike Einstein's view, the particles *do not have a fixed spin*; each particle is assumed to spin *simultaneously in both directions* until a measurement is performed that will force them to take on a fixed spin—a peculiar concept, which is known as the Copenhagen interpretation of quantum theory.[13] It is this interpretation that Einstein found

---

[13] Werner Heisenberg, *Physics and Philosophy: The Revolution in Modern Science* (New York: Harper Torchbooks, 1962).

illogical and aimed to refute. And so did Schrödinger (one of the main creators of quantum theory): to capture the peculiarity of the indeterminate spin state that particles A and B were assumed to be in before the act of measurement, he used a metaphor, the famous Schrödinger's cat. Briefly, he argued that according to the Copenhagen interpretation, until an actual observation is performed, a cat in a sealed opaque box, which also contains radioactive atoms with a chance to decay and spread poison, is both dead and alive at the same time. Namely, the state of existence of the cat before the observation is a mix of two possible outcomes because the status of the cat depends on the status of the radioactive atoms, which, per the Copenhagen interpretation, themselves are in a mixed state of two quantum probabilities, that of the decay outcome, which will kill the cat, plus that of the nondecay, which will preserve the cat. Only after opening the box and observing can the observer actually determine whether the cat is definitely either dead or alive—and that knowledge is of course true only at the moment of observation, not before. So reality, in quantum mechanics, is subjective; it is created only by the act of observation—the moon[14] exists only if we look at it, and a tree falling in a forest makes a sound only if we are there to hear it. Before the observation, the cat doesn't even exist in the Copenhagen interpretation, but its probable existence is expressed mathematically with all opposite qualities simultaneously, such as being both dead and alive (or being both here and there). But according to classical physics, even before opening the box, the cat is in a definite state of existence: it is either only dead or only alive and only at one specific location. So reality, for classical physics (for Galileo, Newton, Einstein) is objective—the moon exists even when we don't look at it, and the tree makes a sound even when no one is there to hear it. (The nature of reality, a much contested topic, has not been settled yet, if ever, for we lack a theory of everything.)

So, according to the quantum view, the spin direction of each particle is fixed by the very act of measurement. For example, if Alice measures that particle A spins clockwise, *then and only then the spin of particle A becomes fixed* (contrary to Einstein's view, for which A would have been spinning clockwise since its creation); and, equally important, *then and only then the spin of particle B becomes fixed, too,* and it is counterclockwise (also contrary to Einstein's view for which B would have been spinning counterclockwise also since its creation). In general, measuring a property of particle A instantly forces a certain fixed property for particle B, even though particle B is not measured directly. This view, which is really the phenomenon of quantum entanglement, appeared absurd to Einstein because it meant, he argued, that the measurement of the spin of particle A affects and fixes *instantaneously* the spin of particle B, even when such measurement is

---

[14] Abraham Pais, *Subtle Is the Lord: The Science and the Life of Albert Einstein* (New York: Oxford University Press, 2005), 5.

performed while the particles are light-years apart and across the universe. This instantaneous "spooky action at a distance," as nicknamed by Einstein, was not required by his analysis since according to it the particles had presumably fixed spins since their creation. How can such instantaneous influence exist, Einstein thought. How is it that measuring a property of one particle instantly affects and fixes an earlier indeterminate property of another particle? How is it that the very moment that the spin of particle A is measured, A communicates instantly how it spins to particle B so that particle B can spin opposite (to conserve momentum)? It is a strange type of communication that occurs faster than the speed of light, in fact instantaneously, and appears to violate one of the main principles of relativity, that information cannot be transferred with a speed faster than that of light because it would violate causality. This bizarreness caused Einstein to believe that quantum theory was not a complete theory of nature despite its immense technological success, which today includes computers, cell phones, and so on. Analogously, Newtonian gravity is abundantly useful in plotting orbits to the moon or building skyscrapers, but its philosophical essence is really incorrect. Observing solely particle A would not in any way influence particle B, which is spatially separated from A, Einstein thought.

But he was wrong. These opposite views appeared for a while to be part of the unverifiable realm of metaphysics. But in 1964 physicist John Bell (1928–1990) found a way to convert each point of view into an experimentally testable calculation (a *number* that can be measured), which is known as Bell's inequality.[15] The experimental verdict found Einstein's view false and favored the spooky action at a distance of quantum entanglement! Indeed, by measuring the properties of particle A, we instantly affect the properties of particle B regardless of how far apart they are. And so, generally speaking, by measuring a property of one particle in a system, what we actually measure is a property of the whole system— which includes us, the observer, too—or, more precisely, of the entire universe. The universe is indeed an indivisible entangled whole. In the Copenhagen interpretation the observer is really part of what he observes—there is a mutual influence between observer and what is being observed. However, in classical physics the observer is thought of as an outsider separated from what he observes—there is no influence at all between observer and what is being observed. The classical-physics view of an observer is therefore like someone watching a movie—if the movie is nature, then an observer eating popcorn and drinking soda while watching has no influence on the movie plot (on nature). Whereas the quantum view of an observer is like someone being in the movie—his actions are *part* of the plot.

---

[15] J. J. Sakurai, *Modern Quantum Mechanics* (Menlo Park, CA: The Benjamin/Cummings Publishing Company, 1985), 226–229.

Of course, this constant and instant interconnectivity between things in the universe, this quantum entanglement, exists not just when we curious observers of nature exercise our "free will" (chapter 14) and decide to make a measurement but is an intrinsic property of nature. For just as in an act of measurement, for which we observers cause willingly particles to interact in order to satisfy our inquisitive mind—for example, photons are shined upon electrons to see where they are and how they spin—particles in nature are in constant interaction anyway (without us having to cause it at will), as if nature itself is constantly self-measuring (self-observing). Now, with self-measuring in mind, we have an additional reason to reinforce a previous conclusion, that, not only are the phenomena observed to occur discontinuously (as a result of the very act of observation, as argued in chapter 7), but the phenomena *themselves* might occur discontinuously even when we are not observing, for *nature* is *self-observing*.

The whole universe experiences the phenomenon of quantum entanglement. If two particles have a chance to interact initially (that is, to become entangled like particles A and B that were created from the decay of the same mother particle), they continue to interact (they remain entangled) even when they are later separated. With this in mind, the entire universe may be considered an entangled whole (where everything in it is in constant and instant interaction with everything else, a perfect Parmenidean whole), for initially, according to the big bang, the entire universe was a mere microscopic size, possibly just a single point, where certainly everything was in close interaction and thus entangled with everything else, and so must then continue to be so today even with everything so far apart. The cosmic interconnectivity of mathematical nature anticipated by the Pythagoreans is now taking a concrete form through quantum entanglement.

Let's justify quantum entanglement. Recall the wave-particle duality (chapter 6). A particle is both a particle (a localized entity) and a wave (an extended entity). The probable existence of particle A is represented by a wave function, and so is the probable existence of particle B. Although of a more complex shape, a wave function is like a straight line, it extends throughout the universe. Since both particles are—in the sense of the wave function—simultaneously everywhere, they are also simultaneously together at any one location; thus, they are constantly entangled.

In concluding this section, I would like to emphasize that information that travels faster than the speed of light is still impossible, as stated in the theory of relativity. That is, while Alice's measurements of the properties of particle A influence instantaneously the properties of particle B, still information, that is, what Alice knows concerning the properties of either particle A or B, cannot be communicated to Bob faster than the speed of light; each person's knowledge can be communicated to one another at best at the speed of light, say, by a radio signal. Only then can Alice and Bob verify the remarkable correlation between the properties of particles A and B due to the phenomenon of quantum entanglement. Before such communication, the outcome of Bob's measurement concerning the spin of particle B would appear to him as random, as dictated by the laws of quantum probability, even when Bob does his measurements after Alice has done hers.

So as attested indirectly by the motion, change, and plurality of everyday experience—when these, of course, are investigated rationally by the human mind—the universe is indeed an indivisible whole. But there is a hypothesis that such universal oneness was once truly absolute.

The Absolute Oneness

The ultimate example of Parmenidean oneness, wholeness, and completeness, as properties of the universe, comes perhaps from the cosmological model of the big bang. It speculates that all matter and energy, all of space and time, the absolute wholeness of the universe of today, was once, about 13.8 billion years ago, contained in just a singular point. This primordial point, we must emphasize, was not within the universe; this one point *was* the universe, the *whole* universe; infinitely small, hot, and dense, containing a single type of particle and obeying one grand law—the absoluteness of oneness, with no sense of location or flow of time, with neither here nor there, neither now nor before or after: the whole universe then (and "there") was but one place of space and one moment of time, a single space-time point.

Unfortunately, the properties of the universe at this hypothetical primordial singular state cannot be described even in principle by our current physical theories. At this singularity—when the universe's size and age are both identically zero—all our physics equations break down; they are meaningless. Could this breakdown be an indication that such a singular state of the universe is really an impossibility, a Not-Being? If so, the universe might have been very small but not point-like. But was it born?

Unborn and Imperishable

Parmenides's philosophical worldview is, so he says, presented to him as a revelation by a goddess and is described in his poem *On Nature*. The main parts of the poem are the "Way of Truth"[16] (which discusses his philosophy) and the "Way

---

[16] Graham, *Texts of Early Greek Philosophy*, 211–219.

of Opinion"[17] (which, among other things, discusses the philosophies of other philosophers). His primary goal was not so much to create a specific physical theory that would explain particular phenomena of nature but rather a theory attempting a logical explanation of existence itself: how can something be? It just is, he reasoned, for there is no such thing as nothing. Nature is unborn and imperishable. That which exists can neither be created from nonexistence nor obliterated into nonexistence. If the universe had a beginning, it would mean that it once did not exist—for if it existed it could not begin. But if the universe did not exist, it would have been Not-Being, and so again it could not begin (for Being cannot come from Not-Being). So the only way to explain why the universe is, is to assume that what is, has always been, unborn, without a beginning.

Now, on the one hand, his view of an unborn nature means that nature has not been caused; it does not have a primary cause. On the other hand, the opposite idea is that nature has been caused by a primary cause. This latter view is in a sense antiscientific since the premise of science is comprehensibility. But a primary cause cannot be understood—if it could, we would know what caused the "primary" cause; hence, the "primary" cause would not have been really primary. Conversely, an unborn nature seems, at least at a first glance, to be more in accordance with the scientific premise, because something unborn/uncreated does not require a primary cause (an explanation) of why it exists—for it has always been.

That said, the notion of an unborn (uncreated, uncaused) nature (or, analogously, an imperishable nature having no ultimate purpose) must be examined with more caution. For it does not exclude the possibility of a god coexisting in the whole—as is, in fact, the case of the Parmenidean "Way of Truth," according to which the apocalyptic goddess Parmenides, and all the rest of nature, all just are. Moreover, an omnipotent and omniscient god could have made nature appear uncreated to us mere mortals. The point is that science cannot prove or disprove the existence of a god, and therefore such a notion, as Parmenides might have put it in his "Way of Opinion," will always remain a matter of subjective belief. In science we must always begin with an assumed something (a Being)—and if we happen to finally explain such assumed something, we explain it with a new assumed something. Science cannot begin from Not-Being: *there is no scientific explanation of a universe coming to be from nothing!* Why there is something instead of nothing is scientifically[18] unanswerable. Causality in our theories explains only later effects by earlier causes, but it cannot explain the primary cause (the beginning). And as a consequence, there is no way to ever know if there is an ultimate purpose. Even if the truth of the universe is revealed to us,

---

[17] Ibid., 219–33.
[18] Within the context of the scientific method.

the only way we can know that such truth is *absolute* is if we ourselves have absolute abilities—so that we can comprehend the absoluteness in the revealed truth. But we do not. And so again, the interpretation of such hypothetical revelation is subjective.

Among our best cosmological models in science, the big bang does not and cannot answer why there is a universe; it only assumes that there is one (that might have begun or might have always been) and then continues to describe it. But it cannot answer why there is what there is. The prediction of the big bang model, that the age of the universe is 13.8 billion years, is only a *relative* age, namely, that our scientific theories can begin describing the properties of the universe roughly since 13.8 billion years ago. But we emphasize that with regard to what the universe might have been doing before that, we are clueless.

Interestingly, in an effort to avoid the breakdown of the equations at the hypothetical big bang singularity, some cosmological models attempt to model mathematically a self-reliant finite universe with neither space nor time boundaries—that is, an unborn nature having neither beginning nor end.[19] A geometrical analogy of such type of universe is the surface of a sphere: its size is finite and no place on it can be considered more of a beginning than an end, more of a center than an edge, or more special in any way. Of course, once more we emphasize that a mathematically/scientifically unborn universe does not say much about the true origin of the universe—that remains a subjective matter—because the truths of mathematics are restricted by its axioms and the truths of science are restricted by its mathematics *and* its scientific method. Scientific knowledge is provisional.

Lastly, the Parmenidean view is a hopeful philosophy because within its context consciousness is part of Being—I think; therefore, I have consciousness. Hence, consciousness can never become Not-Being, even with the body's apparent death.

Conclusion

After Parmenides, any new natural philosophy would be considered incomplete unless it could address successfully his various conclusions, which, though unconventional, were logical. And as if that by itself was not a formidable task, Parmenides's best student, Zeno, assertively supports his teacher's views by adding to the complexity with his famous paradoxes that question the very nature of plurality, space, time, and the reality of apparent motion.

---

[19] Stephen Hawking, *A Brief History of Time: From the Big Bang to Black Holes* (New York: Bantam Books, 1988), chap. 8.

9

Paradoxes of Nature

Introduction

Through a series of so-called paradoxes, Zeno of Elea (ca. 490–ca. 430 BCE) tried to argue for the astonishing conclusion that motion is impossible and plurality is an illusion. Could he be right? We present four of his most daring paradoxes: the dichotomy, the Achilles, the arrow, and the space, which challenge various views on space, time, and motion, and examine them within the context of modern physics. We also refer briefly to the conclusion of his paradoxes on plurality, which deal with whether there are many things or just one.

There is still no commonly accepted resolution for any of Zeno's paradoxes, a fact that preserves their legacy as the most difficult and long-standing puzzles. Part of the reason for this is the involvement of key notions such as space, time, and matter, of which their true nature is far from known even by the standards of modern physics. The real resolution of the paradoxes might require an even more radical understanding of these notions than the one presently provided by general relativity and quantum theory. Proposed solutions have often aimed to prove that motion is real. Empowered by the uncertainty principle of quantum mechanics, we will argue in favor of Zeno that at best, the phenomenon of motion is experimentally *unverifiable*!

The Dichotomy Paradox

According to Aristotle's account, Zeno said, "Nothing moves because what is traveling must first reach the half-way point before it reaches the end."[1] In order to interpret this quote, we must suppose that space is either infinitely divisible (where space is imagined to be divided to ever smaller fractions) or finitely divisible (where space cannot be divided beyond a fundamental length).

---

[1] Aristotle, *Physics* 239b9–14, trans. Daniel Kolac and Garrett Thomson, *The Longman Standard History of Philosophy* (New York: Pearson, 2005), 33.

Infinitely Divisible Space

The paradox can be interpreted two different ways, both of which are essentially the same. In the first interpretation, the question is this: can a traveler ever start a trip? To begin a trip of a certain distance, a traveler must travel the first half of it, but before he does that he must travel half of the first half, and in fact half of that, ad infinitum. Since there will always exist a smaller first half to be traveled first, Zeno questions whether a traveler can ever even start a trip.

In mathematical language, the traveler will be able to start his trip only if he can first find the smallest fraction (the "last" term) from the following infinite sequence of fractions of the total distance: 1/2, 1/4, 1/8, 1/16, ... But such smallest fraction does not exist; it is indeterminable (in fact, this is what is meant by calling such a sequence of fractions infinite). So the paradox is this: while on the one hand, Zeno's argument, which questions the very ability to even start a trip, is logical; on the other hand, all around us we see things moving. Hence, either Zeno's reasoning is wrong or what we see is false.

In the second interpretation, the paradox can be reformulated in a sort of reverse manner. In such case the question will be: assuming a traveler can somehow start a trip, can he ever finish it? To finish a trip of a certain distance, a traveler must first travel half of it, then half of the remaining distance, then half of the new remaining distance, ad infinitum. Since there will always exist a smaller last half to be traveled last, Zeno questions whether a traveler can ever finish a trip.

First "Answer"

First, note that getting up and walking, as Antisthenes the Cynic[2] did after listening to Zeno's presentation and thinking that a practical demonstration is stronger than any verbal argument, is not at all a refutation of Zeno's paradoxes of motion, because Zeno does not deny *apparent* motion; he questions its truth. The natural philosophers were well aware of the deceptiveness in apparent reality; what we see happening is not necessarily happening the way we see it.

An "Answer" Based on Simple Mathematics

With the first interpretation in mind, to start the trip, the traveler must first figure out the smallest fraction of the total distance, that is, the "last" term of

---

[2] Elias, *Commentary on Aristotle's Categories* 109.20–22. Or see Richard D. McKirahan, *Philosophy Before Socrates* (Indianapolis: Hackett, 2010), 182 (Kindle ed.).

the infinite sequence of numbers 1/2, 1/4, 1/8, 1/16, ... Only then he will know where to step first and begin the trip. But such a term is indeterminable. After infinite subdivisions of the total distance, the "last" term of the sequence is indeed infinitesimally small and *approaches* zero, though it is not *exactly* zero: there will always exist a smaller first half to be traveled first. Now since such a term approaches zero, we might want to approximate it to exactly zero. But with such approximation the traveler will step first where he is already, the beginning. This might be interpreted to mean that the trip cannot start; thus, motion is impossible. Nonetheless, this is not necessarily the best conclusion since it is reached only after our convenient approximation of the "last" term with the number zero. Since the actual value of the "last" term is indeterminable, a better conclusion would be that, indeterminable must also be the status of the trip (whether the trip can ever begin). Thus, the notion of motion is, to say the least, ambiguous. The same conclusion is obtained through similar arguments applied to the second interpretation of the paradox.

An "Answer" Based on Modern Mathematics

Often an answer to the paradox is sought through calculus. Suppose the trip distance is 1 meter. Then, as per interpretation two, a traveler first travels half of the trip distance, that is, 1/2 of a meter, then half of the remaining distance, that is, an additional 1/4 of a meter, then half of the new remaining distance, that is, an additional 1/8 of a meter, ad infinitum. To find out if the traveler covers the trip distance of the 1 meter, we must add all the segments traveled by him, that is, 1/2 + 1/4 + 1/8 + 1/16 +. . . . Because the sum of this infinite geometric series converges on 1, some argue that the distance traveled by the traveler after infinite steps is 1 meter; thus, he has moved and the paradox is resolved.

But this argument has a flaw hidden in the details of calculus. To be able to do calculus (i.e., calculate series sums like the one in hand), irrational numbers must be approximated with rational. And there exist infinitely many irrational numbers along any space distance. For example, between the point zero (the beginning of the trip distance) and the point of 1 meter (the end of the trip distance), there are infinitely many irrational numbers—such as $\sqrt{2} - 1 = 0.414213562 \ldots$, or half of it, or one third of it, and so on—that must be approximated with rational numbers before any sum is calculated. For example, approximated to four decimal places, the rounded-off value of $\sqrt{2} - 1 = 0.4142$. Zeno, however, seems to tacitly question these very axioms and approximations that are required in mathematics to make the series convergent to a practical and calculable answer. This is because nature, he would claim, does not have to behave according to the result of such convenient and ambitious human approximations. "Mathematics can never tell us what *is*, but only what *would* be if [this or that axiom is assumed]."[3] Furthermore, some argue that the convergence method does not address the paradox because it does not explain how an infinite number of tasks (going from the first half of the distance to half of the remaining, etc., ad infinitum) can be carried out in finite time.

Now, in the first interpretation of the paradox, motion can't start, and in the second interpretation, motion can't end. These contradictory results hint that the very premise of the paradox, the infinite divisibility of space, might be flawed. Epicurus (we'll see in chapter 13) assumed that space is finitely divisible, that it has a quantum nature! That resolved the paradox. But to be successful, he had to reimagine space and time to be even more radical than the way Einstein did.

The Achilles and the Tortoise Paradox

"In a race the faster runner can never overtake the slower. Since the faster runner must first reach the point from which the slower runner departed, the slower runner must always hold a lead" (Aristotle's account of Zeno).[4]

The paradox says that in a race between, say, fast Achilles and a slow tortoise, initially separated by some distance, Achilles can never overtake the tortoise because before he achieves that he must first reach the starting location of the tortoise. But by the time he arrives there, the tortoise will have had the chance to move to a new location forward; and by the time he arrives at the tortoise's new location, the tortoise will move farther forward to another new location, ad infinitum. Therefore, despite that faster Achilles will be constantly approaching the slower tortoise, still there will always exist some small and ever-decreasing distance separating them (though not necessarily in fractions of half, as in the dichotomy paradox). This is a paradox because, despite Zeno's argument that a faster runner cannot overtake a slower one is logical, fast runners apparently do overtake slower ones. Is Zeno's reasoning flawed, or are our senses false?

This paradox is basically the same as that of dichotomy, so everything mentioned earlier applies here. In the dichotomy paradox, the first interpretation (for which a traveler cannot start his trip) seems to deny motion more directly than the second interpretation (for which a traveler is assumed to move, although he cannot ever finish his trip). In the Achilles paradox, Achilles and the tortoise are assumed to be moving, but motion seems to not work in the conventional way,

[3] Bertrand Russell, *The History of Western Philosophy* (New York: Simon & Schuster, 1945), 124.
[4] Aristotle, *Physics* 239b14–20, trans. Demetris Nicolaides. Or see Daniel W. Graham, *The Texts of Early Greek Philosophy: The Complete Fragments and Selected Testimonies of the Major Presocratics* (Cambridge: Cambridge University Press, 2010), 261 (text 18).

for the faster Achilles cannot overtake the slower tortoise. In the arrow paradox, Zeno is even more audacious, for he directly denies motion by any interpretation. Reconstructing it reads as follows.[5]

The Arrow Paradox

An arrow is at rest whenever it is in a space equal to itself. A launched arrow goes through its flight one instant at a time. Since the arrow is in a space equal to itself each instant of the flight (just as it is when it is at rest), then the arrow must be at rest at each such instant as well. Since it is at rest at any one instant, it must be at rest for the entire duration of the flight. Hence, the flight is apparent, not real; the arrow does not move.

This is a paradox since its conclusion is based on a logical argument that contradicts the apparent reality of sense perception according to which a flying arrow changes positions each moment of its flight and thus apparently moves. Again, is Zeno's logic flawed or are our senses?

I believe the formulation of the arrow paradox must have been triggered by a simple observation, that an object at rest occupies a space equal to its own size. A book, for example, resting on a desk occupies a space exactly equal to its own size. That said, I am not implying that such observation validates Zeno's conclusion that an arrow in apparent flight does not move. But could he be right? Could it be true that an apparently flying arrow is really motionless? Using quantum theory, I will argue that, at best, it is not verifiable whether the arrow is moving or not. Motion, in general, is an ambiguous concept.

Motion Is Ambiguous

While motion is part of apparent reality and is also the very premise of important theories of physics, on a fundamental level (i.e., concerning the motion of microscopic particles, to say the least) motion is not an experimentally verifiable phenomenon, ever! Therefore, motion is essentially a postulate inferred from sense-perceived experiences, but its truth is actually ambiguous. This is so because inherent in the Heisenberg uncertainty principle observations are disconnected, discrete events; consecutive observations have time and space gaps—we can observe only discontinuously (as seen in chapter 7). The concept

---

[5] Ibid., 239b30–33. Or see Graham, *Texts of Early Greek Philosophy*, 261 (text 19); ibid., 239b5–9. Or see Graham, *Texts of Early Greek Philosophy*, 261 (text 20); Diogenes Laërtius 9.72. Or see Graham, *Texts of Early Greek Philosophy*, 261 (text 21).

of continuity in observation must be dismissed. It is a false habit of the mind created by the observations of daily phenomena—as of an arrow in flight (although, as explained in the section titled "Observations Are Disconnected Events" in chapter 7 and as will be reemphasized later, the arrow's apparent continuity of motion is an illusion due to its large mass that makes the time and space gaps between consecutive observations undetectably small). Now without the ability to observe continuously, motion not only *is observed* to be discontinuous but the very notion of motion *itself* becomes ambiguous. How so?

Motion occurs when during a time interval a particle (e.g., an electron) changes positions with regard to some observer; a particle should be now here and later there in order to say that it moved. But since nature does not allow us to keep a particle under continuous observation and follow it in a path, and also since a particle is identical to all other particles of the same family (for example, all electrons are identical), it is impossible to determine whether, say, an electron observed in one position has moved there from another, or whether it is really one and the same electron with that observed in the previous position, regardless of their proximity. Since observations are disconnected, discrete events— with time (and space) gaps in between, during (and within) which we don't know what a particle might be doing—subsequent observations of identical particles might in fact be observations of two different particles belonging in the same family and not observations of one and the same particle that might have moved from one position (that of the first observation) to the next (that of a subsequent observation). Without the ability to determine experimentally whether a particle has changed position, its motion—and motion in general—is a questionable concept.

In summary, (1) without the ability to keep a particle under continuous observation, (2) it is impossible to establish experimentally its identity, and therefore (3) it is also impossible to experimentally verify that it has moved.

Reinforcing this conclusion is the fact that, when we observe a microscopic particle, all we see through a microscope is just a flash of light, and somewhere within it is the particle. But where exactly within it the particle is each moment of time, and whether it is at rest or in motion, are all indeterminable; while we do detect a particle, we detect it neither at rest nor in motion. Hence, indeed, neither immobility nor motion is experimentally verifiable. Motion is ambiguous.

But if we accept it does occur, motion is *quantum*! It occurs through "quantum jumps," a foundational notion in modern quantum mechanics: an elementary particle *has moved* from here to there *but without moving through any of the points in between*! In a *sketchy* analogy, you see a pawn on a square of a chessboard. Then you close your eyes for a bit. Upon opening your eyes, you see *a* pawn on another square. Although you don't know if your observations are of the same pawn, and although you haven't seen a pawn moving, you may still say, *the* pawn *has moved* to a new square *without moving*. Epicurus (see chapter 13) taught this very quantum nature of motion! He arrived at it by countering Aristotle's arguments which were *against* quantum motion. Paraphrased by classicist David J. Furley, Aristotle (in his *Physics Z*) thought if quantum motion were true, then "a thing may have moved without ever moving; a thing may be simultaneously at rest and moving; and motion may be composed of non-motions."[6] Amazingly, trying to capture the peculiar consequences of the Heisenberg uncertainty principle concerning motion, physicist J. Robert Oppenheimer (1904–1967) said something similar: "If we ask, for instance, whether the position of the electron remains the same, we must say 'no'; if we ask whether the electron's position changes with time, we must say 'no'; if we ask whether the electron is at rest, we must say 'no'; if we ask whether it is in motion, we must say 'no.'"[7]

Now, since motion is ambiguous for microscopic particles, then in a stringent sense it must be ambiguous for arrows, too, for arrows are made of microscopic particles. Observing an arrow in flight moving continuously does not prove (in the strictest sense of the word) that one and the same arrow has endured and moved, simply because there is no proof that any of its component microscopic particles have endured and moved. Besides, quantum theory (hence the uncertainty principle) is true for both the world of the large and the small. It is only for practical purposes that the world of the large is assumed to behave according to classical physics—for which objects appear to endure and move in definite traceable paths—because the consequences of the uncertainty principle for large objects are undetectably small, although not zero.

Lastly, in Einstein's block universe, *all* arrow events—where the arrow is each moment of time, that is, all its space-time points—exist in the continuum always, *without* the arrow to have to change position each moment.

Well, then, how do we explain everyday apparent motion and in fact the apparent continuity of apparent motion for any object, such as an arrow in flight, Achilles chasing a tortoise, or anyone taking a trip?

Cinematography and Apparent Motion

We can explain them with cinematography. In an analogy, consider a series of identical and disconnected red lightbulbs, closely spaced along the arched outline of George Washington Bridge. Now imagine it is nighttime and that the first

---

[6] David J. Furley, *Two Studies in the Greek Atomists* (Princeton, NJ: Princeton University Press, 1967), 119.
[7] Robert J. Oppenheimer, *Science and the Common Understanding* (New York: Simon & Schuster, 1954), 40.

lightbulb in the series is turned on briefly for a few seconds and then off forever; after a brief time gap, lasting a minuscule fraction of a second, the second light-bulb is turned on and off the same way, then the third, and so on, until each lightbulb is turned on and off in this sequential manner. The events, the on-off turnings of each lightbulb, are (1) identical (in the sense that an observer sees the same red light) and (2) disconnected: the space gaps are the distances between the bulbs; the time gaps are the minuscule fractions of a second. Furthermore, (3) assuming that the space and time gaps of these events are small enough, to a distant observer, this phenomenon will appear as if *one red object* (the first light-bulb) has *moved* and has moved *continuously* along the outline of the bridge, when in fact no object has. Motion in this case is an illusion of the senses created by observing a series of identical and appropriately disconnected events. In particular, the first two facts, (1) and (2), create the illusion of motion, and requirement (3) creates the illusion of continuity of motion.

The way the red light appears to be moving is similar to the way an arrow in flight appears to be moving. In each case, apparent motion and apparent conti-nuity of motion are in reality the result of (1) a chain of identical observations (of an apparently identical red light or arrow), which (2) are also disconnected (for the arrow, this is due to the uncertainty principle) with (3) undetectably small space and time gaps in between (for the arrow, this is due to its large mass). Specifically, (1) and (2) create the illusion of motion, of one and the same object, the red light or the arrow, and (3) creates the illusion of *continuity* of motion.

But to refine the analogy, we must add this: unlike the case of the bridge for which several identical lightbulbs are assumed to preexist along its outline, for an arrow in flight we cannot assume several identical arrows to preexist along its apparent path; only that, each one of our *observations* is of an apparently iden-tical arrow; though it is uncertain whether our observations are absolutely of one and the same arrow, for, as we learned in chapter 7, it is impossible to establish experimentally the identity of microscopic particles, and since arrows are made up of such particles, it is also impossible to prove unambiguously whether our observations are actually of one and the same arrow (that is, of an arrow com-posed of the *same* microscopic particles at each location of its apparent flight)—a fact that makes motion an ambiguous notion for any object, microscopic or mac-roscopic. The most we can say is that, at subsequent observations, the observed arrows have similar bulk properties, similar general *form* or *organization* (just as the Heraclitean river); it is this general organization that seems to endure, at least during some interval of time and within some region of space, creating the illu-sion of a permanent thing (e.g., an arrow) in motion.

While on the one hand, all around us certain things appear to endure (appear as *permanent things* at least for some time and within some region of space), and whenever they appear to move, there appears to be continuity in their motion, on the other hand, neither permanency in things nor motion is an experimentally demonstrable idea. Thus, motion is, to say the least, an ambiguous concept, a result to be expected because of the very definition of motion, which requires that permanent things exist so they can move: motion occurs when during a time interval a *particle* (a *thing* in general) changes positions relatively to an observer; but to refer to a particle and define its motion, the particle must *remain the same for the duration of its motion*; motion cannot be defined if a particle does not remain the same for a period of time. Now for Heraclitus and modern physics there are no identifiable particles, no permanent things, only events. And without an enduring thing, without the ability to establish the sameness of a thing at two different moments, motion remains an ambiguous concept. While ambiguous, can it nonetheless be a practically useful way to explain the phenomena?

Adequacy Versus Truth

The answer to the previous questions is sometimes. Causality in classical physics is deterministic: a single cause produces a single effect, and both cause and effect are precisely determinable at least in principle. In quantum theory, causality is probabilistic: causes and effects are expressed in terms of a probability; this is actually the reason it is impossible to determine whether the observations of two identical particles might be observations of one and the same particle, for we cannot causally connect these observations with deterministic (absolute) accuracy. Still, to make sense of the phenomena from the point of view of quantum theory, we often assume certain causal chains of events. For example, an electron here collides with a photon and recoils there (as if the electron endures). Thus, while neither a cause nor an effect is certain, and motions are untraceable, still the assumption of a certain chain of events and of motion is often an adequate way to model a practical explanation. Supposing that particles endure, we have previously argued that they are constantly moving.

But while motion may be an *adequate* and useful concept in devising a certain practical explanation of nature—especially so for macroscopic objects such as arrows, cars, and planes—as a *true* property of nature, it is, to say the least, an ambiguous concept, for it lacks the support of experimental confirmation from the microscopic constituents of matter that make up all macroscopic objects. Therefore, the merit of modeling an object (an electron or an arrow) as moving is a practical necessity of everyday life, not a confirmable truth. It should also be pointed out that even practicality cannot be applied consistently (especially so in the microscopic world).

Before quantum theory, and within the context of classical physics, concepts such as position, velocity, and motion (in general) were intuitive, self-evident, and could be used in a definitive way to characterize an object; an object moves with a specific velocity, it is now passing from here and will later go there. However, after quantum theory all these concepts became counterintuitive and could not be used in the same definitive way to describe the behavior of microscopic particles; such particles have neither definite position nor velocity nor a path of motion—as seen, this ambiguity of motion first came up in the natural philosophy of Zeno. A better way to describe the behavior of a particle (and in general the phenomena), then, is not through motion but through the probabilities calculated from the wave functions of quantum theory. It is the concept of probability that is the fundamental (intrinsic) property of matter and not properties such as position, velocity, and motion. Within this context, as initially argued in chapter 6, a particle is truly a mathematical form. Could this mean something concerning change and motion in nature?

A Compromise: Yes to Change, No to Motion

In the view of Zeno, the arrow itself exists but does not move, and there is no change. On the other hand, in the view of Heraclitus, the arrow as a permanent thing does not exist (only events exist), but constant change and constant motion do; only the *organization* of the arrow exists and endures at least for some time. Could these antithetical views be reconciled by modern physics? Well, we can observe an electron (or an arrow) here now and an electron (or an arrow) there later. So obviously we can experimentally confirm that there is a certain change of events (at least in what we observe and where and when we observe it; that is, our observations are of different phenomena, electrons or arrows, at different places and times); but we cannot experimentally confirm that anything has moved. So the compromise might just be this: that constant changes do exist in nature (as Heraclitus posited), but motion does not (as Zeno theorized). But do these changes occur in a passive, playground-like space, or are space, time, and matter somehow related?

The Space Paradox

"If everything that exists is in some space, then that space, too, will exist in some other space, ad infinitum" (Aristotle's account of Zeno).[8] We may reconstruct this quote as follows.

---

[8] Aristotle, *Physics* 209a23–25, trans. Demetris Nicolaides. Or see Graham, *Texts of Early Greek Philosophy*, 263 (text 24).

(1) Things that exist do so in some space.

(2) Space exists (for if it didn't [1] wouldn't hold and things could not exist).

(3) Since space exists and everything that exists is in some space, then space, too, must exist in some other space, ad infinitum.

With this paradox Zeno seems to be arguing that requiring space, that is, void (as the atomists do, by treating it as a sort of a playground to put things in), is as problematic as denying space (as Parmenides does). And that, if we shouldn't completely deny space, we also shouldn't treat space as a playground—as if space is supposed to exist independently of the objects merely for the objects' sake, namely, for them to exist in it.

As discussed in chapters 6 and 7, the special and general theories of relativity replaced the immutable playground-like space of Newtonian physics (for which space and time are absolute) with the malleable space-time continuum (for which space and time are relative). For relativity, things do not just exist in a passive space with time flowing steadily in the background. Instead, space, time, and matter are complexly intertwined—with astonishing effects such as length contraction, time dilation, relativity of simultaneity, and space distortions (the latter being true only in the theory of general relativity).

So Zeno's space paradox is a paradox because while, on the one hand, his argument against a passive playground-like space is logical, on the other hand, it contradicts sense perception of exactly that kind of space. With the theory of relativity in mind, the space paradox may be considered resolved.

The peculiarity of a playground-like space implied by the space paradox is appreciated further when we try to construct a similar-type time paradox (though this is not one of Zeno's paradoxes). For example, if everything that exists does so for some time, then that time, too, will exist for some other time, ad infinitum. This time paradox argues against an absolute (Newtonian) time, flowing the same way for everyone while things happen.

Last, in his effort to show that a nature made up by many things is as problematic and contradictory as the Parmenidean oneness, Zeno devised several other paradoxes. Based on them, he concluded that if in nature there are many things, they must simultaneously be (a) infinitely small and (b) infinitely huge and (c) finitely many and (d) infinitely many.[9] We will not cover these paradoxes here, however, briefly, while the simultaneous existence of opposite qualities (such as of a thing being both large and small) appears as a contradiction, still don't dismiss it easily; recall the Copenhagen interpretation and also read chapter 11.

---

[9] Simplicius, *Physics* 140.34–141.8. Or see Graham, *Texts of Early Greek Philosophy*, 255 (text 7); ibid., 140.27–34. Or see Graham, *Texts of Early Greek Philosophy*, p. 259 (text 13).

Conclusion

Zeno's paradoxes challenge our views of space, time, and matter. Are these notions somehow connected? Einstein thought that, yes, space-time is a continuum in constant and intricate interaction with matter. Is the nature of matter continuous—spread everywhere and also infinitely divisible for which matter can be cut into ever smaller pieces—as Empedocles and Anaxagoras thought? Or is the nature of matter atomic—and so finitely divisible, for which matter cannot be cut beyond some fundamental pieces that are spread unconnectedly because they are surrounded by void—as Democritus taught? Are there also atoms of space and of time, as Epicurus taught? Is there just one primary substance of matter, or are there many? Although all (Empedocles, Anaxagoras, Democritus, and Epicurus) had speculated plurality in the number of primary materials, each had a unique take on it.

10

The Chemistry of Love and Strife

Introduction

Empedocles (ca. 495–ca. 435 BCE) managed to reconcile the antinomies between the Heraclitean becoming (the constant change) and the Parmenidean
Being (the constancy) by introducing four unchangeable primary substances of
matter: earth, water, air, and fire, later called elements, and two types of forces,
love and strife. Change was produced when the opposite action of the forces
mixed and separated the unchangeable elements in many different ways, an idea
in basic agreement with modern chemistry or, more fundamentally, with the
standard model of particle physics. Moreover, the cosmological cycles, described
in his unique cosmology, could have addressed successfully the deceptively
simple question, Why is the sky dark at night?, before the cosmology of the big
bang had it figured out in the twentieth century.

Elements and Forces

Unlike Thales, who taught that the primary material can transform and change
its nature (for example, water can become ice), Empedocles held (as did
Anaximenes) that the nature of a primary material must always remain the
same, like the Parmenidean Being. But with a single primary material of unchangeable nature, he could not account for the observed material diversity of
the world. Thus, he postulated four such materials, the elements, which were
uncreated and imperishable—neither born out of nothing nor perishable into
nothing. His choice for these elements was wise because with them he could
explain the three phases of matter: the element earth could account for the
solid phase, water for the liquid, air for the gaseous. Furthermore, through fire
he could account for light. Now, not only do the elements not change into one
another; they do not change at all. But that did not matter. Because Empedocles
explained nature's enormous diversity by imagining love (the force) mixing the
elements with one another and strife (the other force) separating them from
each other, in infinitely numerous proportions and combinations, forming
composite objects or dismantling them. For example, love can mix earth and
water to produce mud, but strife can separate the earth and water from mud.

Hence, love causes attraction of unlike elements (thus, in a sense, per Aristotle, indirectly it also causes repulsion of things that are alike). And strife causes repulsion of unlike elements (thus, in a sense, indirectly it also causes attraction of those that are alike).

Empedocles explained the unique properties of objects in terms of the proportions of the elements they contain. A hot object, for example, contained more fire than a cold object. And a wet object contained more water than a dry one. Thus, the quantitative difference of the various materials present in an object determines the qualitative difference between objects.

Birth and growth occur while the elements mix, as in a blooming flower, and decay and death occur while the elements separate, as in a shriveling flower. Coming to be (the birth, the generation of something) occurs simply from a mixture of things (the elements) that already exist, not from Not-Being (nothingness)—that is, there is no absolute birth. And perishing (the death of something) occurs simply from a separation into things that also already exist, not into Not-Being—that is, there is no absolute death. That there isn't absolute birth or absolute death is, of course, part of the Parmenidean thesis and is also accepted by Anaxagoras and Democritus.

Like the elements, the forces were corporeal, uncreated, unchangeable, and imperishable. But it was *their* motion through the elements that caused the elements to move, too—either pushing them together to mix or pulling them apart to separate. Hence, forces were the source of motion and consequently of change.

Force, in natural philosophy, appears for the first with Empedocles, who interprets nature in terms of matter and forces. Matter and force, however, became popular with Newton's work: first, with his three laws of motion, and second, with his law of the universal force of gravity. According to his second law of motion, for example, the cause of motion is a force: you pull an object and the object moves. Also matter can produce a force: the sun produces gravity, or an electron the electric force. Nonetheless, while the matter-force interpretation of nature is still immensely practical, it began fading away in twentieth-century physics: forces, in modern physics, gradually became no longer essential. This is a topic to be revisited in chapter 12. Force, in particular, an action-at-a-distance type (as is Newton's force of gravity), we will see, remarkably, was never required in the atomic theory of Democritus.

Empedocles and the Standard Model

Empedocles's idea of forces mixing and separating a fixed number of primary materials is in fundamental agreement with the standard model of particle physics. Whereas Empedocles proposed two forces and four primary elements (renamed "particles" by Lederman),[1] the standard model considers three fundamental forces—the electromagnetic, the nuclear weak, and nuclear strong (recall gravity is not part of the standard model). It also considers twelve types of particles of matter—the six quarks and six leptons (or actually more by more detailed considerations). Of course, unlike Empedocles's elements, in modern physics, quarks and leptons are changeable; they transform to energy or from one material particle into another, although again, these transformations occur by obeying conservation laws, and so in essence something is still unchangeable (e.g., the total electric charge is the same before and after a transformation). Still quarks and leptons are brought together by the forces in a multitude of combinations and proportions to form atomic nuclei, atoms, molecules, flowers, and in general all the plethora of small and large objects, animate and inanimate, similar and dissimilar; but the forces can also break down larger objects into smaller ones.

In Empedocles's chemistry every object is made by a unique mixing proportion of the elements—for example, a bone, he says, is two parts earth, two parts water, and four parts fire (though the sources do not explain how he derived that). (His Pythagorean influence[2] is evident in his use of numerical ratios in mixing processes, just as the Pythagoreans used ratios to describe music, the motion of heavenly bodies, even in their failed attempt to find a ratio for the √2.) Analogously, in modern chemistry every chemical compound is made by a fixed proportion of the chemical atoms—for example, a water molecule, H_2O, is always made of two atoms of hydrogen and one of oxygen. Of course, modern chemistry can be analyzed even more fundamentally within the context of the quarks and leptons of particle physics and still preserve Empedocles's notion of fixed proportions. That is, H_2O, for example, is really a fixed mixture of two protons (one from each hydrogen nucleus) plus eight more protons as well as eight neutrons (from the oxygen nucleus) plus two electrons (one from each hydrogen) plus eight more electrons (from the oxygen). Now, electrons belong in the lepton family of particles and, thus, they are fundamental (they are not made of other types of particles), but protons and neutrons are not fundamental: a proton is made from three quarks, two up and one down ("up" and "down" are quark names); a neutron is made from one up and two down quarks. In addition, quarks and leptons are kept together or pushed apart via the continual exchange of force particles, the photons and gluons, in our example. Analogously,

---

[1] Leon Lederman and Dick Teresi, *The God Particle: If the Universe Is the Answer, What Is the Question?* (Boston: Houghton Mifflin, 1993), 340.

[2] Telauges, the son of Pythagoras and Theano, was Empedocles's teacher. See Greek book *Προσωκρατικοί (Presocratics)*, vol. 6 (Athens, Greece: Kaktos, 1999), 229, https://www.kaktos.gr/000967 (accessed July 15, 2019).

in Empedocles's theory the fixed proportions of the elements are achieved via the constant competition of love and strife.

Empedocles was interested not only in the composition and changes of individual objects but also of the world as a whole.

The Cycles of the World

The structure of the cosmos is spherical for Empedocles, and the changes in it occur without an ultimate purpose or divine intervention (the latter is also the view of the atomists). Instead, nature is ruled by chance and necessity (by potentiality and actuality, in Aristotelian philosophy): namely, only some outcomes are possible, but which ones actually occur is completely the result of chance. Interestingly, this is the meaning of probability in quantum mechanics.[3] For example, only some energies are possible for the hydrogen atom, but the energy (the necessity, or actuality) which is actually observed at any one time is completely the result of chance—the probability of each potentiality is calculated from the wave functions.

Nature in the cosmology of Empedocles has always existed (thus, it has an infinite past time), for, like Parmenides, he believes *ex nihilo nihil fit* (nothing comes from nothing, in Latin). Thus, nature has no beginning or end. It goes through everlasting cycles of growth and decay, gradually and continuously, through four basic periods.[4] In the first period of the cycle, love dominates totally but temporarily, mixing the elements completely. In the second period, strife begins its influence, and so there is a gradual transition to partial mixing and separating. In the third period, strife dominates totally but also temporarily, separating the elements from each other completely, so each, in its pure form, occupies a different region of space: one region of the universe is occupied only by earth, another only by water, another only by air, and the last one only by fire. In the fourth period, love makes its gradual comeback, and so again there is a partial mixing and separating of the elements. Life (the evolution of plants and animals) and nature in general as we know it (with the sun, planets, stars) are happening during the second and fourth periods. The state of our cosmos is temporary for Empedocles, and it is gradually being succeeded by another.

Now, for a universe with no beginning or end, with "first" and "last" to have no absolute meaning, why should one thing (or concept) be more primary

---

[3] Werner Heisenberg, *Physics and Philosophy: The Revolution in Modern Science* (New York: Harper Torchbooks, 1962), 27–28, 122.

[4] Simplicius, *Physics* 158.1–159.4. Or see Daniel W. Graham, *The Texts of Early Greek Philosophy: The Complete Fragments and Selected Testimonies of the Major Presocratics* (Cambridge: Cambridge University Press, 2010), 251 (text 41).

(fundamental) than another? "Again, if everything is created from four things [earth, water, air, fire, or quarks and leptons] and resolved into them, why should we say that these are the elements [the primary, the fundamental entities] of things [of the composite objects or of the emergent properties] rather than the reverse—that other things [the composite objects] are the elements [primary, fundamental entities] of these? For one gives birth to another continually, and they interchange their colors and their entire natures [properties] throughout the whole of time."[5] With this Epicurean reasoning in mind, nature might be a two-way street, nonhierarchical: subtle sameness, which we are in search of to construct a theory of everything, might be as significant and fundamental of a concept as its opposite, perceptible diversity. Generally, reductionism— understanding (constructing) the universe from the bottom up (microscopically to macroscopically)—might just be as valid a philosophy as its opposite, emergentism, understanding (deconstructing) the universe from the top down.

Interestingly, if we are not myopic in our comparisons, these four periods have several similarities with modern cosmological models of the universe.

Cycles in Modern Cosmology

According to the big bang model, initially everything was completely mixed together, space, time, matter, and energy (like Empedocles's first period). Life as we know it was then impossible because the universe was tiny and superhot, without stars or planets, just a super-dense mixture of tiny particles. The universe has since then been evolving, reaching its astronomical size and diverse state of today, with galaxies, stars, planets, and life (as in Empedocles's second period). Now, if, as speculated by various big bang models, the universe is "open," it will continue to expand forever, increasing its size so much that ultimately everything in it will be completely separated (as in Empedocles's third period). It will then be a cold, lifeless universe without planets or stars, only isolated tiny particles. But if, as also speculated by other models, the universe is "closed," then after it goes through a third period (a state of maximum, though not necessarily complete, separation of everything in it, during which stars might fade out and die), it will stop expanding and will begin contracting, resulting again in life-bearing partial mixing and separation (as in his fourth period). The fourth period is much like the second, for as matter is brought together in a shrinking universe, the particles coalesce again to form countless light-giving stars and life-sustaining planets to orbit them. But in a "closed" universe the contraction will continue until the

---

[5] Lucretius, *On the Nature of the Universe* 1.764–769, trans. R. E. Latham (London: Penguin Books, 2005), 28.

crushing force of gravity ultimately collapses the universe in on itself, and brings once more everything completely together (the first period all over), causing a "big crunch" (the opposite of the big bang). If the universe were to reverse and begin to close, would the second law of thermodynamics reverse, too? As is, the second law is supposed to dictate the "arrow of time," the familiar directionality (flow) of time from past, present, future, where things (flowers, people, stars, the whole universe) *all* tend toward greater disorder (entropy) and basically grow old. If, in a closing universe, however, the second law were to reverse (i.e., the *order*, not the disorder, of things were to increase), the arrow of time would reverse, too, in which case we might feel getting *younger* as in the reversed cosmos of Plato's cosmological myth.[6]

We are not sure if we live in an ever-changing universe going through endless cycles of big bangs, expansions, contractions, and big crunches. Still, we could describe rather accurately the main events in the universe and when they occurred by starting from the "first" moment of the big bang until now.

Cosmic Calendar

In modern cosmology all events in the universe span 13.8 billion years in time. To gain a perspective of such time vastness, we often employ a cosmic calendar. It is a metaphor by which 13.8 billion years, the estimated age of the universe since the big bang, are compressed into just one calendar year—about a month per 1 billion years. The initial bang, the big bang, is supposed to have happened at precisely midnight, 00:00:00 (which, in the 24-hour time notation, is the 0th hour, 0th minute, 0th second) on January 1, causing the expansion of the universe. What caused the bang is still unknown, although it is speculated to have been a kind of repulsive gravity that is predicted from the equations of general relativity. But what is known is that this expansion has been happening ever since and up until now, the last moment of December 31 at 24:00:00, increasing the size of the universe from an unimaginably small size initially, possibly point-like, to today's immensity, of about 93 billion light-years across. (Whether the universe is finitely or infinitely big or old, are questions that in truth have not been settled yet, if ever, because cosmology is a developing field. If we adhere to the philosophy that Being can neither come from Not-Being nor become Not-Being, then the universe has always been and will always be. Also it can be boundless without being infinite, as is the geometry of a sphere discussed in chapter 8.) Anyway, what banged (expanded, stretched)? Space-time did and still does. Within a

---

[6] Plato, *Statesman*, 268–274e.

minuscule period of time after the initial bang, possibly by a mere 10^{-36} second, the universe underwent an immense faster-than-light[7] expansion, a *big* bang, an idea known as cosmic inflation. In the blink of an eye it expanded by a factor of 10^{30}![8]

By about 14 minutes (380,000 years) after the big bang, at 00:14:00 on January 1, the universe expanded, became less dense, and cooled significantly and as a result became transparent to light (as a clear-air day is to visible light), allowing for the first time the "afterglow" of the big bang, the oldest observable light formally known as the cosmic microwave background, to travel freely through space and time, from there and then to here and now, and to be seen today (by radio telescopes) coming from every direction in the universe—a triumphal verification of the big bang cosmological model. Earlier than the first 14 minutes, the young universe was very dense and hot and thus opaque to light—as a foggy day is to visible light—thus, light could not travel far. January 1, at 00:14:00, is also the instant that the simplest and lightest of the chemical atoms, hydrogen, first formed when a relatively cooler universe allowed electrons and protons to capture each other via the electric force.

Stars and galaxies began to form by around February 1 (about a billion years after the big bang) from matter pulled together by gravity. The era before stars (and starlight) is called the Cosmic Dark Ages. Stars shine because of nuclear fusion, the process via which light nuclei combine to form heavier ones, converting mass into energy and releasing light. Nuclei heavier than iron, including silver and gold, are synthesized via fusion when a supergiant star (more massive than the sun) becomes a supernova—dies violently in a cosmic explosion, producing as much light as a galaxy of stars!

Its death is also life's birth! For gradually after millions or billions of years, a supernova's scattered debris, an interstellar cloud of gas and dust, collapse again under the crushing force of gravity and grow into a new star with its orbiting planets that may also develop life. A perfect example is our own solar system. It was born much, much later, around September 3 (about 4.5 billion years ago) from the gravitational collapse of a massive interstellar cloud that was composed from the atoms that were synthesized earlier in the universe, including the heavy atoms made in the stars. Thus, earth and everything on it, including us, are all made of these ancient atoms—if you are wearing a gold ring, you are in a sense actually wearing a portion of a star, for your jewelry's

---

[7] In relativity, material objects can't travel faster than light through space, but space itself can expand at speeds greater than c.

[8] Jeffrey Bennett, Megan Donahue, Nicholas Schneider, and Mark Void, *The Essential Cosmic Perspective*, 7th ed. (Boston: Pearson, 2013), 450.

atoms were once manufactured in a supernova-destined star! Even more impressive, in the words of the great Carl Sagan, we are all "star stuff"! In other words, most of the atoms we are made of were once made inside stars that lived and died millions or billions of years before we or our own solar system were even born.

Primitive microscopic life forms were thriving on earth by September 29 (3.5 billion years ago), so the first type of life must have evolved much earlier than that. On December 30 (65 million years ago), an asteroid collided with the earth and caused the extinction of many species, including the dinosaurs. But that was a good day for primates because that's when they started to evolve. *Homo sapiens*, which are primates, evolved on the last hour of the last day of the cosmic calendar, December 31 at 23:52 (only 8 minutes ago, 200,000 years ago). And at different moments during the last minute of the last day various other significant events occurred. Humans painted fine cave art 1 minute ago at 23:59 (30,000 years ago). They domesticated plants and other animals and gave birth to civilization 23 seconds ago at 23:59:37 (about 10,000 years ago).

Recorded history, which preceded the construction of the pyramids by a few centuries, began only 11 seconds ago at 23:59:49 (about 5,000 years ago), and the birth of Greek natural philosophy occurred just 6 seconds ago, at 23:59:54 (2,600 years ago with Thales). Our innovative Internet was implemented about 0.08 seconds ago at 23:59:59:92 (in the 1980s), and a 20-year-old reader of this book was born only 0.05 seconds ago at 23:59:59:95. If wisdom is, as the wise say, acquired with time, then human wisdom is only infinitesimal, not at all like that of the cosmos, infinitely universal.

What will a second such cosmic calendar be like for the universe? Will the universe continue to expand? Will it stop and begin to contract? We are not sure. While the cosmic microwave background and Hubble's law constitute two of the most significant experimental verifications of the universe's expansion, an experimental verification concerning the universe's fate (if a particular one does exist) is yet to be found. Experiments are important because they verify or falsify a scientific hypothesis. Empedocles is known to have done an experiment, possibly the first in the history of science.

It's Experiment Time

While air was the primary substance of matter in the philosophy of Anaximenes, still it was not accepted as a real corporeal substance for two reasons: (1) it is invisible and (2) because other objects appear that could be placed in air or move through it. So, within the context of these reasons, air was thought, at least by the Pythagoreans, to be really the void. But using a clepsydra (a device to lift and transfer liquids), Empedocles overturned such belief by experimentally proving that air is indeed a material substance.[9]

Submerge a straw (which is much like a clepsydra) in a glass of water. Water flows into the straw through its bottom opening and fills it as high as is the water level in the glass. But if before you submerge the straw you first cover its top opening with your finger, no water (or, actually, very little) will flow into the straw. This happens, Empedocles argued, because some invisible material, which is already trapped in the straw, presses on the water (through the bottom opening of the straw) and keeps it out; water, in this case, cannot move through this material. This material is air. Only when you uncover the top opening can water once again flow in the straw. For in this case the air in the straw escapes through the top opening, and so an equal volume of water flows in to take its place. (Incidentally, why water or air or any object can move will occupy the mind of the atomists, as will be seen in chapter 12.) In conclusion, since it is not always true that an object can move through air or be placed in it, air must be a material substance, regardless of its invisibility. Empedocles's reasoning is correct.

Why Is the Sky Dark at Night?

The apparently simple question, which is known as Olbers's paradox, puzzled astronomer Heinrich Olbers (1758–1840), who was probably the first to have asked it, and all others until the cosmology of the big bang was used to resolve it. But it wouldn't have puzzled Empedocles.

Assume (as Olbers and most thinkers of his time did) that the universe is static (neither expanding nor contracting), infinitely old, infinitely big, and filled uniformly with (sun-like) stars throughout (the infinity of space and time). If you imagine gazing in some, *any* direction, your line of sight, Olbers thought, will eventually intercept a star.[10] Thus, the sky must always be as bright as the sun. But it's not. It's a paradox because the argument is logical, but its conclusion contradicts the evidence. The resolution of the paradox is given by the cosmological model of the big bang.

---

[9] Aristotle, *On Youth, Old Age, Life, Death, and Respiration*, 473b9–474a6. Or see Graham, *Texts of Early Greek Philosophy*, 387 (text 127).
[10] That is, (a) starlight from infinitely distant stars, Olbers thought, will *eventually* reach our eyes. But that's arguable because infiniteness is not an actuality (eventuality); it's only a potentiality. Thus, alternatively, (b) such starlight will potentially be traveling *indefinitely* (during the infiniteness of time), without *ever* actually traversing the infiniteness of space that separates us in order to be seen, in which case the paradox is no longer a paradox. Besides, if we are to treat infiniteness as an actuality, as in (a), then (c) an infinitely distant star is actually infinitesimally small, a *true* point, a *zero*-size star, a non-existent source of light; thus, the view toward bunches of such stars would be of darkness (not of light), as are the dark patches of the sky at night. But Olbers didn't think of these alternatives.

The night sky is dark primarily (1) because when we look out in space, we look back to the time of the Cosmic Dark Ages when there were no stars—during January in the cosmic calendar. So the darkness we "see" is the space we "see" before stars existed. The stars we see began to form from February onward and shine in contrast to that January darkness. Moreover, stars don't last forever; they evolve and die out. Olbers had incorrectly assumed that stars existed forever. By contrast, although Empedocles's universe is infinitely old like that of Olbers, the cycles of Empedocles's cosmology allow ingeniously for periods without stars—of Cosmic Dark Ages; thus, the paradox does not hold.

There are other reasons, too. The night sky is dark, (2) because the afterglow of the big bang, which fills all of space, is microwave, not visible, light. (3) Because of the expansion of the universe, the light emitted by the receding galaxies is redder and fainter than if they remained still. (4) Because the young universe was slightly lumpy, its density varied from place to place. Thus, only the denser regions evolved to be the galaxies, whereas the less dense became the practically empty space between galaxies.

An equally important question is, why is the day sky bright? It is bright (and blue) only for that part of the earth that faces the sun, during daytime. Because sunlight then, which is scattered by the atmosphere, is seen coming from every direction of the sky. If there were no atmosphere to scatter the sunlight (as is, for example, on the moon), both "day" (when the earth faces the sun) and "night" sky (when the earth faces away from the sun) would have been dark, always—although the patches of the sky that have the sun (and the other stars) would be bright.

The Origin and Evolution of the Species

In his effort to understand the origin of the species and their adaptation to their environment, Empedocles, like Anaximander, conceived of an evolutionary theory by natural selection. In the beginning a chancy mix of the "immortal"[11] (permanent, unchangeable) elements created all imaginable "mortal"[12] (temporary) organic "forms, a wonder to behold."[13] These, though, were just parts, from humans, animals, and plants. And so "many heads sprouted without necks, and arms wandered bare and bereft of shoulders, and eyes strayed up and down

---

[11] Simplicius, *On the Heavens* 529.1–17, trans. Graham, *Texts of Early Greek Philosophy*, 361 (text 51).

[12] Ibid.

[13] Ibid., trans. Bertrand Russell, *The History of Western Philosophy* (New York: Simon & Schuster, 1945), 54.

in need of foreheads."[14] That is, until love mixed them in countless ways more, so that the species of plants and animals formed. But only the fit survived; the unfit died. When a human head, for example, combined with a human body, the creature acquired a fitting form and survived, Empedocles thought; but when a human head combined with an ox body, he continued, the creature was unfit and died. Chancy material combinations and natural selection (that is, survival of the fittest and adaptation) are important aspects in both Empedocles's and modern theories of biological evolution.

Conclusion

Empedocles's pluralistic philosophy was a crucial turn away from the monistic philosophies we have discussed so far (i.e., those that considered water, the apeiron, or air as the only primary substance of matter, or the philosophy of Parmenides about oneness), for it paved the way for the most successful ancient pluralistic philosophy, the atomic theory of Leucippus and Democritus. Their theory required myriad particles: the atoms. But before the theory of atoms, natural philosophy had to go through yet another theory of remarkable originality; four primary substances of matter for Empedocles, but infinitely many for the nous of Anaxagoras and everything is in everything.

[14] Ibid., 586.12, 587.1–2, trans. John Burnet, *Early Greek Philosophy* (London: A & C Black, 1920), chap. 7 (frag. 57).

11

In Everything Is Everything

Introduction

"Nous [the mind] set everything in order"[1]; thus, it has the ability to understand nature rationally, Anaxagoras (ca. 500–ca. 428 BCE) proposed. Order though, according to him, is not achieved through the consideration of just one primary substance or even four but through a countless number of them, including things such as gold, copper, water, air, fire, wheat, hair, blood, bones, and in general all other existing substances. However, unlike Empedocles's four elements, which are pure, Anaxagoras's substances are not; "in everything there is a portion of everything,"[2] a notion as bizarre as two of the most popular interpretations of quantum theory, the Copenhagen and the many-worlds.

In Everything There Is a Portion of Everything

Every piece of substance, however large or small (at any magnification), contains some portion of everything—portions can be large but infinitesimally small, too, because for Anaxagoras matter is infinitely cuttable. Hence, no one substance is more fundamental (that is, smaller, simpler, purer) than any other. But "each thing is most manifestly those things of which it has the most."[3] A piece of gold, for example, contains gold as well as everything else—copper, wheat, hair—but appears as a distinct golden object because its gold content is the greatest. This does not mean, however, that this golden object contains the substances in pure form, side by side, separated, and identifiable, and the amount of pure gold in it happens to be more. No! To the contrary, no matter how small a bit we may cut from such a golden object, it will still contain a portion of everything—it will never be pure gold. Generally, *no* part of *any* object is ever pure, for an object

---

[1] Simplicius, *Physics* 164.24–25, 156.13–157.4, 176.34–177.6. Or see Daniel W. Graham, *The Texts of Early Greek Philosophy: The Complete Fragments and Selected Testimonies of the Major Presocratics* (Cambridge: Cambridge University Press, 2010), 291 (text 31).

[2] Ibid., 164.23–24, trans. G. E. R. Lloyd, *Early Greek Science: Thales to Aristotle* (New York: W. W. Norton & Company, 1970), 44.

[3] Ibid., 27.2–23. See Gregory Vlastos, *Studies in Greek Philosophy: Volume 1 The Presocratics* (Princeton, NJ: Princeton University Press, 1993), 319.

(or any bit from it) can't both be pure and obey "in everything is everything." Therefore, despite that this is a golden object, *every part* (bit, length scale) of the object is also *simultaneously* watery, woody, milky, bloody, bony, hairy, and every other material; *and*, in every such part, every material preserves its fixed proportion in relation to all others. But not just that—it gets stranger.

An object has not only a portion of each type of substance but also a portion of all *opposite* qualities. As with the substances, these qualities are not to be assumed to be side by side in an object or separated somehow, as if, say, an object has its right side wet and its left dry. Rather, "Things in this one cosmos are not separated from one another, nor are they split apart with an axe, neither the hot from the cold nor the cold from the hot."[4] So every part of an object (e.g., every location in an object), or generally, an object *is all the qualities simultaneously* and in a fixed proportion relatively to one another. For example, something hot is to some degree also cold. Or white snow, Anaxagoras argued, is to some degree simultaneously black, too—a statement of the same unusual meaning as Schrödinger's cat being simultaneously both dead and alive.

Anaxagoras and the Copenhagen Interpretation

So for Anaxagoras an object is simultaneously hot, cold, wet, dry, hard, soft, sweet, sour, black, white, bright, dark, dense, rare, dead, alive, spinning clockwise, spinning counterclockwise, and all other opposite qualities. This is a peculiar interpretation of nature, for before we observe an object, the most we can say about the state of its existence is that it is a mix of all possible outcomes—of all opposite qualities simultaneously, though each with a different degree (portion) of contribution. Only after we observe the object can we describe it in a specific way, in terms of "those things of which it has the most," say, as golden, yellow, cold, heavy, and dry.

Remarkably, such an interpretation is similar to the most popular interpretation of quantum theory, the Copenhagen view. According to it, before an observation, something (an electron, Schrödinger's cat) is all opposite qualities (potential outcomes) simultaneously too, with each outcome described by its own quantum probability to actually occur. Recall how before an observation Schrödinger's cat is simultaneously both dead and alive (or how an electron spins simultaneously both clockwise and counterclockwise). And each of these potential outcomes has its own probability to actually happen. Only after we observe, the Copenhagen interpretation states, can we determine whether

⁴ Simplicius, *Physics* 176.29, 175.12–14. Or see Richard D. McKirahan, *Philosophy before Socrates* (Indianapolis: Hackett, 2010), 194 (Kindle ed.).

the cat is definitely either dead or alive (or whether the electron spins definitely in the one or the other direction), and in general, whether an object is, as Anaxagoras states, definitely golden, yellow, cold, heavy, and dry. If the idea of portion in Anaxagoras's theory is roughly associated with the idea of probability in quantum theory, then indeed, "in everything [a system of interest] there is a portion [is described by the quantum probabilities] of everything [of every possible outcome]."

Now the reason Anaxagoras required that various portions of all qualities had to coexist simultaneously everywhere within an object and at all times is that he wanted to remain in accordance with the Parmenidean thesis, that Not-Being does not generate Being, and that Being does not become Not-Being. Something must always exist if it is to be observed, the thesis says. That is, if a quality were not already present everywhere in an object always, it could not have come to be later; because if it did come to be later, it would mean that Being could be generated from Not-Being, but that's impossible. Hence, a hot object, for example, has to contain simultaneously both hotness *and* coldness everywhere within it and always, though in different portions. For if a hot object did not contain coldness, too, coldness would have been Not-Being (at least for that object), and therefore it could have never come to be (coldness could have never become a reality, a part of Being)—it would then be impossible for the hot object to be cooled down.

This concept has a certain similarity with the Copenhagen interpretation but also a certain difference. Concerning the similarity, the reason we may observe the cat to be alive (or the electron to spin clockwise) is that the cat's (or the electron's) state of existence before the observation is a mix of all possible outcomes (including opposite ones), that is, a mix that includes a portion (the quantum probability of occurrence) of the alive quality (or the clockwise spin) *together* with a portion of the dead quality (or the counterclockwise spin). In quantum theory this mix state is expressed mathematically. And the outcome with the highest probability (portion, in the language of Anaxagoras) is the one most likely to be observed.

Analogously, the reason we may observe, say, hotness, in Anaxagoras's view of our previous example, is that the object's state of existence before the observation is a mix that includes a portion of hotness and a portion of coldness, but with the hotness portion being the highest.

But Anaxagoras is even bolder than the Copenhagen interpretation, a fact that brings me to their difference. He insists that the notion of the simultaneous existence of all qualities (opposite ones, too) is true all the time, even after an observation. Hence, in Anaxagoras's view, the cat is still both dead and alive (or the electron still spins in both directions, or the object is both hot and cold) even after we observe the cat to be only alive (or the electron to spin only clockwise, or the object to be only hot). But in the Copenhagen view, after an observation the cat is only alive (or the electron spins only clockwise, or the object is only hot). Thus, although we observe only one quality, and so the cat appears only alive (or the electron is detected to spin only clockwise, or the object to be only hot), for Anaxagoras the other qualities never cease to exist; he insists on this because he does not want to violate the Parmenidean thesis that if a quality ceased to exist, it would mean that that part of Being became Not-Being. Now, can Anaxagoras be somehow right on this, too? Can the cat somehow be both dead and alive even after we observe it to be only alive?

Anaxagoras and the Many-Worlds Interpretation

Fascinatingly, yes! According to the second-most popular interpretation of quantum theory, the many-worlds view, even though *we* observe the cat to be alive (or the electron to spin clockwise), in *another universe* (world, reality) the cat is dead (or the electron spins counterclockwise)! That is, an outcome that is possible but does not occur in our universe still occurs in another universe. In general, every outcome that could have occurred in our present reality (universe) but did not branches off as an alternative reality (it gets realized) in a parallel (i.e., separate) universe; each parallel universe thus has its own unique reality that consists of events that could have happened in our universe but did not.[5]

Hence, the many-worlds view is in closer agreement than the Copenhagen view with both Anaxagoras's theory as well as the Parmenidean thesis. For, Parmenidean Being (being everything that there is) can easily be interpreted to include every possible outcome of an observation. Then, in the view of many-worlds, it is not only before an observation that all possible outcomes (even the opposite ones) coexist in a mix and are part of Being (as is also required by the view of Anaxagoras), but all such outcomes, in a sense, continue to coexist and are thus still part of Being even after an observation (also as required by the view of Anaxagoras); for each possible outcome occurs in its own parallel universe even when such outcome is not observed to occur in our own universe. Whereas on the other hand, in the view of the Copenhagen approach, though before an observation all possible outcomes coexist in a mix and are part of Being (as also required by the view of Anaxagoras), after an observation only what is observed to occur continues to exist (to be part of Being), and what is not observed no longer exists, as if part of what once existed, part of Being, became Not-Being (a situation in clear violation of both the view of Anaxagoras and the thesis of Parmenides). With the Parmenidean thesis in mind, one might then say that

---

[5] We may think of *the* universe to be made of parallel universes.

the many-worlds interpretation of quantum theory is more accurate than the Copenhagen.

Fractal Forerunner

Anaxagoras is the forerunner of the idea of fractals.[6] His philosophy of "in everything there is a portion of everything" together with "an object is those things of which it has the most" implies that if we could zoom in on some particular object, say a golden object, *at any length scale* (at any magnification), we would always observe the same repeating pattern (structure). It would be so because all materials (milk, honey, gold, etc.) coexist simultaneously *and* in the same proportional manner *throughout* the golden object. For example, in a golden object the material gold is the same proportion more than all other materials at all length scales. In general, Anaxagoras's theory of matter entails that any portion of an object (however small) looks the same with its bigger self—like the Matryoshka dolls. This universal self-similarity—of some particular pattern persisting at all length scales—is of course the defining characteristic of fractals.

An example of a fractal is a coastline. It has roughly the same shape, however large or small a coastline segment is. Fractals are ubiquitous, in art, pure mathematics (e.g., Koch snowflakes), biology (e.g., certain plant leaves, brain[7]), physics (e.g., critical phase transitions), and cosmology.[8] For a fractal (or hierarchical) universe, galaxies are imagined to form clusters, which form superclusters, which in turn form super-superclusters, and so on, by preserving some initial cluster pattern (figures 11.1 and 11.2). A fractal universe is so far a cosmological hypothesis. But the fractal nature of a critical phase transition is a fact.

Consider as an example the phase transformation of water, from liquid to gas. The boiling temperature of 100 degrees Celsius occurs at the pressure of 1 atmosphere. The liquid phase coexists with the gaseous phase at these values, with each phase retaining its own unique properties (e.g., water is denser than steam). The boiling temperature rises as the pressure increases. But at the specific temperature of 374 degrees Celsius and the pressure of 218 atmospheres, something very special, *critical*, happens. Then, we no longer have two distinct phases coexisting, only a single fractal-like fluid phase—a *supercritical fluid*. The structure of this supercritical fluid phase fluctuates continuously more violently than ever before,

---

[6] Petar V. Grujic, "The Concept of Fractal Cosmos: I. Anaxagoras' Cosmology," *Serb. Astron. J.* no 163 (2001): 21–34. Found at http://saj.matf.bg.ac.rs/163/pdf/021-034.pdf (accessed July 15, 2019).

[7] Sean Carroll, *The Big Picture: On the Origins of Life, Meaning, and the Universe Itself* (New York: Penguin, 2016), 323 (Kindle ed.).

[8] Yurij Baryshev and Pekka Teerikorpi, *Discovery of Cosmic Fractals* (River Edge, NJ: World Scientific, 2002).

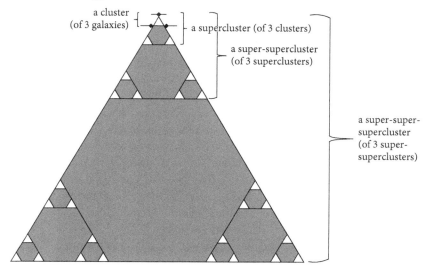

Figure 11.1 Triangular fractal universe in 2 dimensions.

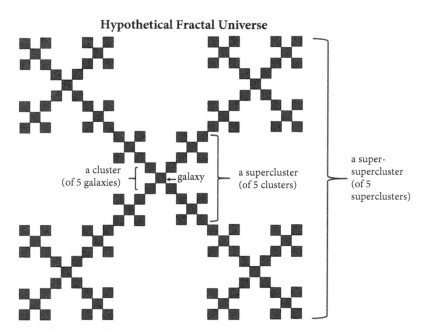

Figure 11.2 Square fractal universe in 2 dimensions.

with bubbles and droplets of *all* different magnitudes blended everywhere, forming, dissolving, and interchanging their natures (phases), and transforming the substance into a perfect fractal—with the same pattern of intermixed bubbles and droplets persisting at all magnifications, as is in Anaxagoras's theory of matter! In a normal phase (solid, liquid, gas) the cooperativity between water molecules is short range; it extends only to nearest neighbors. It is as if you (a molecule) hold hands with two of your friends, who in turn hold hands only with their nearest neighbors/friends, and so on. By contrast, in a supercritical phase, cooperativity is long range; it extends throughout the substance and therefore is much stronger. It is as if everyone is friends, and everyone has as many hands as needed to hold everyone, in all combinations and regardless of each other's distance. This long-range correlation is what preserves the fractal universality in critical phase transitions, even amid the wild fluctuations of the substance. The critical point of a phase transition is the "sweet spot [of perfect fractal universality that separates] boring order and useless chaos."[9]

Anaxagoras's "in everything is everything" has yet another interesting universal application.

The Perfect "In Everything Is Everything"

The ultimate example of "in everything there is a portion of everything" is the big bang singularity, the hypothesis that the primordial state of the universe was once, 13.8 billion years ago, a mere point. "In everything"—in the singularity, which *itself* was everything that existed, the whole universe—"there" was "a portion of everything," matter, energy, space, time, and the laws (or ultimate law) they obey. And the reason the universe is diverse, with planets, stars, people, and plants is that, as Anaxagoras might have explained, there is only a *portion* of everything in everything and "each thing is most manifestly those things of which it has the most."

Now, what remains enigmatic is this: (1) if indeed the notion of "in everything is everything" was true at the singularity, why would it not be true always? And (2) can plurality, all the beautiful and diverse nature of today, unfold from an absolute oneness, the singularity? Both are open questions.

On the first question, Anaxagoras would have answered "in everything is everything" always and everywhere. On the second question, all three pluralists, Anaxagoras, Empedocles, and Democritus, believed that plurality must be absolute; that is, neither could plurality (the many) have come to be from an initial

---

[9] Carroll, *The Big Picture*, 323.

singularity (from what is initially one), nor a singularity (the one) from an initial plurality (from what is initially many). Nevertheless, whether nature is truly monistic or pluralistic is a question that has yet to be answered. The singularity hypothesis is problematic. Will our nous (mind) ever know the nature of nature? Amazingly, Anaxagoras preserves purity in an "impure" universe of "in everything is everything" by assuming that nous, found only in living things, is the only thing pure. However, what is the origin of our ability to reason? Why do humans have an intelligent nous, the most intelligent, in fact, of all the species that we know?

From Walking to Thinking

Anaxagoras believed that the cause of human superiority over animals is the hand. Although other primates walk upright occasionally, only humans walk upright habitually, and as a result only humans have freed two of their limbs permanently and have been using them as hands consistently. This unique trait of ours has been significant for both our biological and intellectual evolution because when Lucy, a species of the genus *Australopithecus*, about 3.2 million years ago "decided" to walk upright more habitually, it meant that two of her four legs were starting to evolve into hands, increasing her potential to use and make tools. Toolmaking stimulates thinking (silent and out loud, thus speech, too), which in turn refines toolmaking, which stimulates further thinking, in a continuous cycle, ultimately advancing both technology and the intellect, and making Lucy's distant relative, the *Homo sapiens* (us), indeed *sapiens* intellectually superior to any other animal (at least on earth). And thus the origin of such superiority might truly be the hands. What actually ignited this development was a purely chance mutation in the spine that allowed our hominid ancestors to stand upright and evolve their forelegs into hands, become environmentally fitter, and get naturally selected further.

Upright posture (and consequently free hands), speech, and a complex brain (nous) are among the most unique traits of the human species and the source of our resourcefulness. The brain is the center of operations. Speech is controlled by the left side of the brain, but the coordination of the movements of our hands comes from the back part of the organ. Such coordination took literally millions of years of evolution to be mastered, during which the brain was driving the hands, which in turn were driving the brain, causing the enhancement of both organs, developing each to its present advanced stage of evolution compared to the hands and brains of other species. Without this type of evolution we would not have been able to make our first tools; stack up stones; build homes, the pyramids, the Parthenon, the Empire State Building; or create cell phones, computers, spaceships, MRIs, or the LASER, but also we would not have been able to pursue other more abstract endeavors, such as religion, philosophy, science, and the arts.

At the same time, however, I wonder if there is a limit to such brain-hands interactive enhancement. Even worse, I wonder, what the risks today are from the plentiful technology made by our hands as a result of the ingenious science conceived by our brains. Do we get to depend more and more on machines to think for us and on pills to save us, risking the weakening of both the mind and the body? If yes, our natural abilities might atrophy and our evolution might be stalled. We might even devolve, for habits have a say on whether we get to evolve. Namely, habits are transmittable culturally through teaching and can still change the environment in complex and subtle ways. And in turn, through the process of natural selection from biological evolution, the environment can influence a species by controlling the direction of its evolution.

The name of Galileo is often associated with the first major conflict between religion and science. He was tried by the Inquisition of the Roman Catholic Church for his support of the heliocentric model, which the Church then considered in contradiction of biblical accounts of an immobile earth in the center of the universe (i.e., the geocentric model). But it was really the science of Anaxagoras that caused such conflict first. He was charged with blasphemy by the Athenian democracy for thinking that "the sun is a fiery stone"[10] and not a god. He was tried and found guilty. Although he was defended by his student Pericles, the famous statesman, still by one account he was exiled, by another he was sentenced to death. Anaxagoras was an original thinker. He is credited with explaining eclipses correctly and for introducing philosophy to the Athenians. When asked why he was born, he replied, "To theorize about the sun, moon, and heaven."[11] When told, "You are deprived of the Athenians," he replied, "No, they are deprived of me."[12]

Conclusion

Whether the nature of matter is infinitely cuttable (without smallest pieces) or finitely cuttable (with ultimate smallest pieces that make up everything) is still

---

[10] Diogenes Laërtius 2.6–15, trans. Demetris Nicolaides. Or see Graham, *Texts of Early Greek Philosophy*, 275 (text 1).
[11] Ibid.
[12] Ibid.

an open question. Anaxagoras held the former, but Leucippus and Democritus held the latter: namely, matter is atomic and thus made up of disconnected, indivisible pieces known as the atoms and surrounded by empty space. What revolutionized science was the atomic theory of matter, an idea that is two and a half millennia old.

12

Atoms of Matter and Energy

Introduction

Perhaps the greatest scientific achievement of antiquity, possibly of all time, was the realization of the atomic nature of matter. "There are but atoms and the void,"[1] Democritus (ca. 460–ca. 370 BCE) proposed. And he understood the great diversity of material objects as complex aggregations of uncuttable atoms, the building blocks of matter, moving in the void, the empty space between them. Leucippus, who flourished between 440 and 430 BCE, invented the atomic theory, and Democritus, a true polymath and a prolific philosopher, developed it extensively. Uncuttable (the actual meaning of *atom* in Greek) are also the modern elementary particles of matter, the quarks and leptons, and although void is a controversial concept still, a kind of void is required to explain nature.

Atoms

Ancient Atoms

Atoms, in the ancient atomic theory, are the uncuttable smallest pieces of matter, disconnected from each other because they are surrounded by void, space devoid of matter that was required to enable the atoms to move. Atoms are invisible, impenetrable, solid (absolutely rigid), indestructible, eternal, unchangeable, unborn (not generated by something else more fundamental), and imperishable (they do not transform into something else more fundamental). Atoms are therefore like many Parmenidean Beings. Unlike the elements of Empedocles, which represented four different types of known materials, or those of Anaxagoras, which represented infinitely many types, all atoms are made from one and the same type of material (although not from any particular one of the everyday, such as water or air). Atoms have no internal structure (they are homogeneous) but differ from each other only as regards their size and shape. Their only behavior is motion. Roaming around in the void of an infinite space there

---

[1] Sextus Empiricus, *Against the Professors* 7.135, trans. Erwin Schrödinger, *Nature and the Greeks and Science and Humanism* (Cambridge: Cambridge University Press, 1996), 89.

exist infinitely many atoms of various shapes: angular, concave, convex, smooth, rough, round, sharp, and so on.

Their motion is perpetual (so, then, is change in accordance to the Heraclitean worldview). Motion continues by itself without requiring a force (or a causal agent in general). The fact that constant motion continues by itself was discovered via experimentation first by Galileo and was later restated by Newton through his first law of motion, also known as the law of inertia ("laziness")—a body remains (because of its laziness) at rest or in uniform motion until affected by a force. Atomic motion is also thought random because space was correctly assumed to be isotropic (having no special location or direction; that is, space has no absolute up, down, right, left, in, out, center, or edge). Hence, when left to themselves (undisturbed), atoms had no reason to move more one way than another—all directions of motion were equally probable. Motion was not explained by Leucippus and Democritus; it was simply postulated to have always been, without a beginning. In fact, atoms and the void were also postulated. Not accounting for the cause of motion received Aristotle's intense criticism,[2] even though it is actually a normal scientific procedure. For we need to remember that postulating the truth of a certain beginning and proceeding from there to understand nature in a causal and rational way is the only way to do science. Science must begin from something (a postulate, an axiom, a primary cause); it cannot begin from nothing (Not-Being). For Democritus the atoms, the void, and motion were part of a primary cause, which by definition requires no cause of its own.

As they move, atoms collide with each other, bounce, and rotate; some hook together (whenever their shapes are complementary) and assemble in a multitude of arrangements, forming all kinds of macroscopic (compound) objects that appear "as water or fire, plant or man,"[3] or unhook and disassemble, deforming (destroying) the objects. As atoms aggregate, objects form and increase in size, and as atoms segregate, objects change form and decrease in size (i.e., they perish). Just as the words "tragedy" and "comedy" are formed when letters (which can be thought of as atoms) from the same alphabet are combined differently, Aristotle had explained, the immense plethora of diverse objects is formed when atoms are arranged in space differently through their motion.[4] In fact, atoms were required by Leucippus and Democritus to explain exactly this

---

[2] Aristotle, too, by the way, didn't account for *his* "natural motion," down toward the earth. He just called it "natural."

[3] Plutarch, *Against Colotes* 1110f–1111a, trans. Daniel W. Graham, *The Texts of Early Greek Philosophy: The Complete Fragments and Selected Testimonies of the Major Presocratics* (Cambridge: Cambridge University Press, 2010), 537(text 28).

[4] Aristotle, *On Generation and Corruption* 315b6–15. Or see Graham, *Texts of Early Greek Philosophy*, 541 (text 41).

diversity in nature: "For from what is truly one a plurality could not come to be, nor from what is truly many a unity, but this is impossible."[5] Remarkably, for Democritus the aggregations and segregations of unseen atoms in the void produced not only our own world (with the earth, moon, sun, planets, stars, plants, fish, animals, and including humans) but also countless others. Earthlike (thus habitable) planets are a commonplace in the universe according to the latest astronomical findings!

Atoms have none of the conventional properties of matter such as color, taste, smell, sound, temperature, or even weight. The indication for this, Democritus thought, is in the fact that various objects are perceived differently by different people. Something sweet to me might be spicy to you. "To some honey tastes sweet to others bitter but it is neither," said Democritus.[6] So he explained the conventional properties in terms of the shape of the atoms and their motion in the void; shape and motion cause unique macroscopic arrangements (in objects, in people) and thus unique conventional (emergent) properties. For example, the atoms of a hard object are more closely packed with less empty space (void) between them than the atoms of a soft object. Now, since the atoms of a soft object have more void to roam around they can be pushed there more easily. Hence, such objects feel squeezable and soft. Metals are made of atoms with hooks that hold them firmly interlocked, but liquids are made of round atoms so they can flow by each other easily. Sweet objects are composed of round good-sized atoms; bitter of round, smooth, crooked, and small; acid of sharp (so they can sting the tongue) and angular in body, bent, fine; oily of fine, round, and small. Even light was made of atoms (particles)—incidentally, Einstein won the Nobel Prize in Physics by interpreting light (pure energy) in terms of discrete particles: the photons. Black, white, red, and yellow were considered primary colors and associated with different shapes and arrangements of atoms.[7] Combinations of these four colors were in turn used to account for all color variations. Democritus worked out a detail theory on sensation. In general he argued that the constant motion of the atoms, which persists even when a composite object is seemingly at rest, causes some of them to be emitted by the object. Flying through the void, these atoms in turn ultimately collide with the atoms of a sense organ and create a unique sensation (a flavor, smell, color, an image).

All changes of the apparent world of sense perception, animate and inanimate, were reduced to the irreducible atoms and their motion in the void. This

[5] Ibid., 324b35–325a6, a23–b5, trans. Graham, *Texts of Early Greek Philosophy*, 529 (text 14).

[6] Sextus Empiricus, *Outlines of Pyrrhonism* 2.63, trans. Demetris Nicolaides. See Greek book Βας. Α. Κύρκος (Vas. A. Kyrkus), *Οι Προσωκρατικοί: Οι Μαρτυρίες και τα Αποσπάσματα τόμος Β* (*The Presocratics: Testimonies and Fragments*, vol. B) (Αθήνα: Εκδόσεις Δημ. Ν. Παπαδήμα, 2007), (Athens: Publications Dem. N. Papadima, 2007), 255.

[7] Graham, *Texts of Early Greek Philosophy*, 579–595.

scientific reductionism is ambitious. In principle, it is also the goal of the modern theory of the elementary particles of matter, the quarks and leptons. But while there are striking similarities between these modern particles of matter and the ancient atoms, there are also serious differences. Only for purposes of comparison let us call the ancient atoms Democritean or D-atoms, and today's building blocks of matter, the quarks and leptons, QL-atoms.

D-Atoms and QL-Atoms: Similarities and Differences

Before we proceed with this comparison, let us first briefly summarize the historic developments in search of the D-atom, the ultimate uncuttable piece of matter. Until about the end of the nineteenth century, chemical atoms, such as hydrogen, carbon, oxygen, and so on, of the periodic table of chemistry, were thought to be the fundamental particles of matter, the D-atoms. But this idea turned out to be incorrect when the structure of the chemical atom was probed further and it was found that it was cuttable, made up of electrons and a nucleus. In 1897 physicist J. J. Thomson (1856–1940) discovered the electron, and in 1911 his student physicist Ernest Rutherford (1871–1937) discovered the nucleus. Electrons (one of the six types of leptons) are still thought indivisible, but atomic nuclei not so; the latter are made from divisible protons and neutrons, though they are themselves composed of indivisible quarks. Six types of quarks were postulated to exist as elementary particles in the 1960s, and all six types had been discovered by the end of the twentieth century. Hence, chemical atoms are not fundamental; they have substructure (they are divisible) and in fact are made of the QL-atoms. Like the shadows, which *were* real but were not the real *objects* in Plato's parable of the cave, chemical atoms *are* real but are not the real *fundamental particles* (the smallest cuts of matter as envisioned by Leucippus and Democritus)—although the name "atom" has been undeservingly stuck on them, for it's really a misnomer (inaccuracy) when used to refer to the chemical elements.

On the other hand, both D- and QL-atoms are fundamental because they are not made from other particles; they are disconnected pieces of matter, indivisible (uncuttable), invisible, and the smallest, and their various combinations make up all material things in the universe. Neither the D-atoms nor the QL-atoms have any of the conventional properties of composite objects. These properties are really a consequence of the collective behavior of the D- and QL-atoms that make up these objects.

D-atoms are unchangeable; they do not transform. But QL-atoms do; they transform from one type of material particle to another and also into and from energy (although they do not transform into something more fundamental). But like matter, energy also comes as discrete bundles, as particles (e.g., photons), and so Leucippus's and Democritus's notion of discreteness as a property of nature is preserved and applies to energy, too. Furthermore, like D-atoms, which are made of the same substance, QL-atoms are made of the same substance, too, mass and energy (which are equivalent as per special relativity). And since D-atoms are indestructible, so is their substance, but so is the substance of QL-atoms, for the total amount of mass-energy in the universe is constant (as per the law of conservation of mass-energy). So the substance of both, the ancient and modern atoms, endures, while nature is constantly changing.

D-atoms have shapes and thus have nonzero size; QL-atoms are considered point-like and thus shapeless and size-less.[8] There is both a challenge and a simplicity associated with each view. In the D-atoms we need to imagine all sorts of complex atomic shapes, but we need not worry about forces. Democritus did not introduce any (a topic to be revisited later). D-atoms, he explained, coalesce into composite objects as a result of their perpetual motion and complementary shapes. Also, composite objects have size since they are made of D-atoms, which themselves have size. On the other hand, QL-atoms lack shape and size, but they still combine. They do so via the exchange of the particles of force (the photons, the W's, the Z's, the gluons, also the gravitons if we find them). Being point-like thus shapeless is in a sense a simplicity, for it means that QL-atoms are internally structureless like the D-atoms. But it is also a complexity, for how can something of zero size, of zero extension in space, have properties such as mass, electric charge, energy, spin, and so on, and, even worse, how can composite objects have size and extension in space when their constituents do not? Within the context of the Parmenidean theory, the "size" question may be restated as follows: how can size come from not-size? That is, how can Being (the nonzero size of macroscopic objects) come to be from Not-Being (from zero-size constituents)? This was impossible for Parmenides. Democritus solved the size challenge by postulating that matter cannot be divisible (cuttable) ad infinitum; it must be finitely divisible with the smallest cuts to be the indivisible, the uncuttable D-atoms of nonzero size. For only then, he thought, could composite objects have size, if they are composed of things that themselves have size. Interestingly, the strings of string theory *do* have size. If these turn out to exist, we may not need to worry about size (Being) coming to be from not-size (Not-Being), an idea that would be pleasing to both Parmenides and Democritus. The size challenge is revisited in the section on "Void or Not?"

On the one hand, D-atoms are postulated to move in order to comply with the apparent world and explain change. On the other hand, since motion is an

---

[8] Although in truth, no experiment has thus far verified that the size of QL-atoms is *absolutely* zero, just that they are so super tiny that they may be treated satisfactorily as point-particles.

ATOMS OF MATTER AND ENERGY   125

ambiguous concept from the modern point of view, QL-atoms are postulated to move only as an adequate way of understanding the phenomena. Furthermore, in quantum theory, material particles (the QL-atoms) are less materialistic in the sense that they no longer have key properties that material particles were once thought to have in order to be called material: they are neither permanent, nor indestructible, nor unchangeable, nor deterministic, nor have they well-defined shapes or trajectories through space and time, and as a consequence nor have they identity and individuality. Thus, QL-atoms are best regarded as events and not as permanent Parmenidean Being-like entities (as are the D-atoms). And the properties of QL-atoms are best described in terms of the quantum probability, a number that expresses only a potential event: for example, which type of QL-atom might be observed, where, and with which properties and behavior.

Are the QL-atoms the smallest cuts of matter and, within this context, thus the ultimate D-atoms? It is generally not thought so. Are the QL-atoms different forms of the same type of universal substance, the same type of particle, and what might that be? While the Higgs boson particles have some qualities required of a universal substance, the standard model that predicts them does not include the most puzzling force in the universe, namely, gravity. Therefore, although useful, any model of nature that does not incorporate gravity is incomplete.

The basic concept of the ancient atomic theory was highly valued by Nobel laureate Richard Feynman. He said: "If, in some cataclysm, all scientific knowledge were to be destroyed, and only one sentence passed on to the next generations of creatures, what statement would contain the most [scientific] information in the fewest words? I believe it is the *atomic hypothesis* (or the atomic *fact* . . .) that *all things are made of atoms—little particles that move around in perpetual motion, attracting each other when they are a little distance apart, but repelling upon being squeezed into one another*. In that one sentence, you will see, there is an *enormous* amount of information about the world, if just a little imagination and thinking are applied."[9] On a related note, in his book *The God Particle* Nobel laureate Leon Lederman graded thousands of scientists (including himself) for their efforts in their quest for a primary substance of the universe. He started with Thales all the way to 1993, the completion date of his book. Democritus received the only A in the class![10]

So as a general *idea*, the D-atoms, the uncuttable discrete and fundamental pieces of matter that everything is made of, are still part of our most advanced theories of nature, for these basic but important properties are properties of the QL-atoms, too—in fact, also of the strings of string theory. But whether the

[9] Richard P. Feynman, *The Feynman Lectures on Physics* (Boston: Addison-Wesley, 1963), 1–2.
[10] Leon Lederman and Dick Teresi, *The God Particle: If the Universe Is the Answer, What Is the Question?* (Boston: Houghton Mifflin, 1993), 340.

QL-atoms (together with the force-carrying particles of the standard model as well as the Higgs boson, a total of sixty-one particles all confirmed to exist), a zoo of other unconfirmed particles of matter and energy (including the graviton, predicted by various other scientific models), or some new particles of *one and the same type* (a much-desired scientific simplicity of Democritean grandeur) that manifest themselves by way of the familiar particles are or will be the truly uncuttable discrete and fundamental pieces of matter remains to be seen. How about the void? Does it exist or not? Is it needed, or can it be avoided?

Void or Not?

The atomists Leucippus and Democritus called an atom a *thing*, Being (what-is), and the void *nothing (not thing)*, Not-Being (what-is-not).[11] And they agreed with the theory of Parmenides (with one interpretation of it, anyway, for which the properties of Being are understood literally) that motion is impossible without the void. But whereas Parmenides denied the existence of the void by considering it Not-Being, the atomists postulated the opposite: Not-Being, the void, exists, for only then, they thought, can motion and change be accounted for. It is the place to put the atoms and enable them to move. For the atomists the void is empty space, and so *in* it there is nothing. But for Parmenides *it, the void, empty space itself*, is nothing, Not-Being; not *in* it there is nothing. The nature of the void has created mind-boggling debates since the time of Parmenides. For if something, for example, the void, is really nothing, how can it exist? How does one define "nothingness"? The answer is not easy.

But first let's summarize Democritus's arguments favoring the void. By accepting the phenomena of motion, change, and diversity to be real, he deduced the void to be real as well, for without the void his impenetrable, indivisible atoms could not move and consequently the phenomena of change and diversity would not occur; but they do occur, so the void must be real. Similarly, by accepting also multiplicity and division of composite objects to be real, he again deduced void to be real, for without it, composite objects could not be divided (cut) into smaller pieces: "division resulted from the presence of void in bodies."[12] As explained further by philosopher and mathematician Bertrand Russell (by paraphrasing Democritus), "When you use a knife to cut an apple, the knife has to find empty places where it can penetrate; if the apple contained no void, it would be infinitely hard and therefore physically indivisible."[13] Here

---

[11] Aristotle, *Metaphysics* 985b4–20. Or see Graham, *Texts of Early Greek Philosophy*, 525 (text 10).
[12] Simplicius, *On the Heavens* 242.15–26, trans. Graham, *Texts of Early Greek Philosophy*, 533 (text 23).
[13] Bertrand Russell, *The History of Western Philosophy* (New York: Simon & Schuster, 1945), 71.

we recall of course that for Democritus divisibility does not continue ad infinitum; it applies only to composite objects and stops at his physically indivisible atoms. So for Democritus both the atoms and the void are real: "thing [atoms] is [exist] no more than not-thing [void]."[14]

Now what does modern physics think of the void? Does it exist or not? Is it a true nothing, the Parmenidean Not-Being, or something else? While "nature abhors a vacuum,"[15] a popular phrase since the Renaissance, yet "nothing works without, well, nothing."[16]

Void?

On the one hand, void is still a useful concept for the understanding of many phenomena. According to quantum theory, electrons in a chemical atom, for example, "move" around their nucleus by keeping their distance from one another as if space between them is empty, devoid of matter—"a regulation against overcrowding"[17] formally known as the Pauli exclusion principle. As a consequence of this principle, the electrons of chemical atoms keep their distance from each other; they do not like to be squeezed together into a small region, so they act as if they were rigid: the closer they get, the faster they move apart—a statement in agreement with the Heisenberg uncertainty principle, for, according to it, the uncertainties in the position and velocity are inversely proportional, so the smaller a particle's region of confinement (the smaller its position uncertainty), the faster its motion to escape such a region (the greater its velocity uncertainty), "almost as if it [the particle] were overcome with claustrophobia," Brian Greene wrote.[18] The exclusion principle explains why chemical atoms are mostly empty space and why macroscopic objects (which are made of chemical atoms) have a degree of rigidity, size, and shape—e.g., the distance that tiny particles keep from each other translates into the size and shape of a macroscopic object. D-atoms are rigid; consequently, in a sense they, too, obey the regulation against overcrowding, for one D-atom cannot occupy the same region of space as another D-atom. Had the exclusion principle not been true, the QL-atoms, which obey it, would not endure as disconnected pieces of matter, thus nuclei would not form, nor would chemical atoms or the molecules of organic chemistry, and

---

[14] Plutarch, *Against Colotes* 1108f–1109a, trans. Graham, *Texts of Early Greek Philosophy*, 527 (text 13).

[15] Isaac Asimov, *Understanding Physics* (xx: Dorset Press, 1988), 7.

[16] Lederman and Teresi, *The God Particle*, 44.

[17] Banesh Hoffman, *The Strange Story of the Quantum* (New York: Dover, 1959), 68.

[18] Brian Greene, *The Elegant Universe: Superstrings, Hidden Dimensions, and the Quest for the Ultimate Theory* (New York: W. W. Norton & Company, 1999), 114.

consequently nor would the matter that living things are made of; generally, *all* matter in such a scenario would collapse into a uniform, undifferentiated, and lifeless state. The diversity in nature is in a sense a consequence of the exclusion principle: diversity is a law of nature!

Or Not?

On the other hand (that is, to be able to explain other phenomena), in the quantum realm the void is not really devoid of matter but a very busy place, seething with all-pervasive fields of energy (e.g., light and gravity waves, even the much-required Higgs field that explains mass—see section "Worlds Without Forces" later), known as vacuum energy. These fields cannot be zero, even in seemingly empty space, because the time-energy uncertainty principle would be in violation (recall the section titled "*Nothing* Comes from Nothing" in chapter 8). And they are actually fluctuating constantly, creating and annihilating pairs of particles with their corresponding antiparticles. These particles, which are called virtual, are not created out of nothing or annihilated into nothing but are made out of energy and return to be energy (the vacuum energy). Unlike real particles, which can be directly observed, virtual particles cannot, even though they can still cause measurable effects on real particles, an indication that "empty" space is really not empty.

Moreover, according to the theory of general relativity, the whole of space is filled by a gravitational field with properties (such as strength) that vary from place to place and from one moment to the next. Einstein explained gravity by assigning *properties* to "empty" space; empty space (and time) is a flexible medium that gets distorted by a mass, and gravity is space's (and time's) distortions. These *properties* are the void, and so in the theory of general relativity the void is not the Parmenidean Not-Being, for Not-Being, a true nothing, is property-less.

Furthermore, an astronomical observation completed in 1998 with the aid of the Hubble Space Telescope found that the expansion of the universe is accelerating—so a galaxy's recession speed measured today is faster than its speed measured yesterday. This accelerated expansion is attributed (although reluctantly) to the existence of dark energy, which is hypothesized to permeate the universe and act as a kind of antigravity by stretching space and causing it to expand at continuously faster speeds. Dark energy, which has not yet been detected, is one of the most puzzling mysteries of the universe. Dark matter is yet another great puzzle: though invisible, for it does not emit light, its existence is inferred indirectly by its gravitational pull on neighboring stars. What makes dark matter invisible, no one knows. "Dark matter is attractive [it attracts stars], dark energy is repulsive [it's hypothesized to push galaxies apart and cause the expansion of the universe]."[19]

Ordinary matter, matter we can see, which makes up flowers, people, planets, stars, and galaxies, is only about 5 percent of the total stuff in the universe. The other 95 percent, which includes dark energy and dark matter, is stuff that we neither see nor know much about, although their subtle presence is deduced in some way.[20] Space, even "empty" space, is a place of constant, frantic activity of virtual particles, light, gravity, dark energy, dark matter, ordinary matter, of Higgs boson particles, possibly strings, and who knows what else. It is certainly not the Parmenidean Not-Being (the *absolute* nothing). Therefore, once more we emphasize that no scientific theory can base its beginning, its first cause/s, its axioms, on absolute nothing, on Not-Being; science must begin from somethingness, and what that might be becomes increasingly more complex. In fact, even Leucippus's and Democritus's nothing (their notion of void) is really not nothing, since from the point of view of modern physics "it [their void] was the carrier for geometry and kinematics, making possible the various arrangements and movements of atoms."[21]

Fascinatingly these atomic arrangements and movements were imagined by Democritus to be carried out without the requirement of a force of, say, gravity, electricity, or magnetism; other than their direct collisions during which there was a physical contact, D-atoms experience no other force! D-atoms have no weight, and they produce no force of gravity.[22] How can this be? How can there be a world without gravity, without forces in general?

Worlds Without Gravity

Weight or gravity (that is, the tendency of objects to fall or the property of heaviness) was not one of the primary characteristics of atoms but a property that was accounted for by Democritus ingeniously through motion, in particular rotational motion.[23]

Though motion is chaotic, Democritus argued, in an infinite space with infinite atoms there is always a chance that the bulk of the atoms of a certain region

[19] Katherine Freese, *The Cosmic Cocktail, Three Parts Dark Matter (Science Essentials)* (Princeton, NJ: Princeton University Press, 2014), 195.

[20] Jeffrey Bennett, Megan Donahue, Nicholas Schneider, and Mark Void, *The Essential Cosmic Perspective*, 7th ed. (Boston: Pearson, 2013), 479.

[21] Werner Heisenberg, *Physics and Philosophy: The Revolution in Modern Science* (New York: Harper Torchbooks, 1962), 40.

[22] Aëtius 1.3.18, S 1.14.1f. Or see Graham, *Texts of Early Greek Philosophy*, 537 (texts 31, 32); Cicero, *On Fate* 20.46. Or see Graham, *Texts of Early Greek Philosophy*, 537 (text 33).

[23] Aëtius 1.4.1–4. Or see Graham, *Texts of Early Greek Philosophy*, 541–545.

move collectively in a preferred direction of motion, rotational in particular, and produce a vortex. The rotational motion of such a vortex, Democritus thought, ultimately causes the bigger atoms (the more massive, the heavier, as we would say today, having gravity in mind) to move toward its center, ultimately forming the earth and the water on it, and the smaller atoms (the lighter) to move toward its outskirts, ultimately forming the air, the sky, and the stars. Because the dynamics of our world system is still rotational (e.g., the sky rotates, relative to us, and so in a sense do the moving clouds), objects made of the bigger atoms, like a rock, still fall, and objects made of the smaller atoms, like steam, smoke, or fire, still rise, Democritus argued. Air, on the other hand, generally does not fall, he thought, because of its rapid revolution, just as water does not spill from a cup when it is rapidly spun around. His analysis was logical as regards observation because the earth, which (for him) is made of the bigger atoms, formed in the center of his vortex, water, made of smaller atoms, is on earth, and air, made of even smaller atoms, is above water and earth.

Now, concerning the dynamics of a vortex, in reality it is the reverse that happens: massive objects tend toward the outskirts of a vortex, and lighter ones toward the center (this, for example, happens in a centrifuge, a device employed to separate different substances). Nonetheless, this error in Democritus's explanation is really a minor point compared to the fact that he managed a reasonably clever explanation of the world only in terms of a basic property that atoms have, namely, their motion in the void. Thus, he saw no need of a force of a weight, of gravity, despite that apples fall as if a force is pulling them through space. The latter, legend says, inspired Newton to conceive his theory of universal gravitation for which gravity was a force, only to be abolished as a force by Einstein's theory of general relativity. How so, and what does quantum theory say about forces in general?

Worlds Without Forces

An apple and the earth, or the sun and the earth, feel a force of attraction from each other, Newton thought, as a consequence of a mysterious action at a distance (not to be confused with the action at a distance of quantum entanglement) that he himself admitted he did not understand. How is gravity transmitted if the interacting objects do not touch each other? How does one body feel the other, how do they communicate, if nothing but empty space exists between them and if nothing specific is really exchanged by them? How does the apple know that the earth is below and that it should fall? Newtonian gravity describes rather accurately how the apple falls, but not why it falls. What's the cause of gravity?

Space is. Einstein discovered through his theory of general relativity that space pushes the apple. More specifically, "Matter [the sun, earth, apple] tells space how to curve [recall the trampoline analogy, chapter 7] and space tells matter [and light, too] how to move,"[24] said physicist John Archibald Wheeler (1911–2008), who summarized general relativity with a single sentence! Light, too, (like matter, the marble on the trampoline) zips through the curved space-time (the distorted trampoline fabric) following a geodesic (the shortest distance, which in a curved space is not a straight line), bends as it passes near massive objects like the sun, and appears to be attracted by them—a phenomenon not predicted at all by Newtonian gravity but understandably so because, unlike the trampoline fabric which is seen pushing the marble, space is not seen directly pushing matter or light; Einstein had to imagine it for us.

Hence, Einstein eliminated the need for an action-at-a-distance-type force by recognizing that the agent that transmits gravity is curved space (in fact, curved time, too, the way it passes) when distorted by matter. Space is no longer the Newtonian passive playground where events unfold but a flexible medium the geometrical shape of which *changes* (gets warped) by matter. The distortion of space (the changing geometry of space) in turn influences an object's motion and feels like gravity. With such a geometrical explanation of gravity, the notion that gravity is a force is abandoned in general relativity. And in the study of the phenomena of gravity, an object's motion through space and time may no longer be regarded as a response to an action-at-a-distance Newtonian force of gravity acting on it, but as a response to the warping—the *geometry* of—space-time caused by the distribution of all other objects around it.

Moreover, as already discussed, according to the standard model, the particles of matter, the QL-atoms, combine with one another via the continual exchange of the particles of force. Recall, for example, that the attractive and repulsive electric force is really a manifestation of intricate particle collisions; even gravity is hypothesized to work via the exchange of gravitons. Matter and force are no longer distinct notions. Instead, forces are really expressions of complicated particle collisions.

And so, as is the view of Democritus, nature can be understood in terms of just particles and their complex collisions—forces were never required in the theory of Democritus and are no longer required in modern physics! Incidentally, although Empedocles's two forces, love and strife, were separate

---

[24] Charles W. Misner, Kip S. Thorne, and John Archibald Wheeler, *Gravitation* (Princeton, NJ: Princeton University Press, 2017), 5.

entities from his four elements, still they were not action-at-a-distance types of force: via their direct contact, they either pushed the elements together to mix or pushed them apart to separate, so they, too, in a way, acted as colliding particles.

Even mass, and consequently weight, is thought to not be a fundamental property of the QL-atoms, rather, a property the QL-atoms acquire through their interactions with the Higgs field. The mass of an object is a measure of its inertia, or of its resistance to a change in its state of motion. The smaller the resistance, the less the mass is. Throwing a baseball is easier than a bowling ball: the baseball has less mass than a bowling ball, or, equivalently, it produces less resistance to our attempt to throw it (and change its state of motion, from rest to moving). Now, the Higgs field pulls on the other particles (e.g., the QL-atoms) as they traverse through it and impedes their motion. It is this resistance that we interpret as mass. In an analogy, stirring a cup of coffee with a spoon is easy, but a cup of honey is not. Honey is a more viscous fluid, and the spoon feels heavier, more massive, as it moves through it. In this analogy, the spoon is a QL-atom and the fluid the Higgs field. Just as fluids with different viscosities cause the spoon that moves through them to feel lighter or heavier by impeding its motion, the standard model imagines that the all-pervasive Higgs field creates an analogous effect on the initially massless QL-atoms as they traverse it, endowing each with a unique mass and slowing them down. It is as if the Higgs field had different viscosity for different-type QL-atoms. Similarly, the force-carrying particles W's and Z's acquire their mass, but photons, feeling no resistance by the Higgs field, remain massless and thus can move with the speed of light. The analogy describes what is known as the Higgs mechanism, which explains why some particles have mass and some not (though it does not explain why they have the actual mass they do). What particular agent gives Higgs bosons their mass is nevertheless still an unknown. The idea that mass may not be a fundamental property was prompted by a few interesting and unresolved questions. Why, for example, is there no pattern in the mass values of the particles, a fact in contrast to other particle properties, such as spin or electric charge? For instance, in some units of measurement, the spin of all QL-atoms is 1/2 and the spin of all the force-carrying particles is 1.

Amazingly, mass is not a fundamental particle property, neither in the standard model of modern physics nor in Democritus's atomic theory! Equally amazing is that in both the modern and the Democritean physics, the nonfundamental property of mass (and the consequent heaviness, weight) is caused (is acquired) by atomic motion—the motion of the QL-atoms through the Higgs field, and the motion of the D-atoms in the vortex!

Particles are by definition discrete entities; thus, their existence implies a certain discontinuity in nature. But is the nature of nature truly discontinuous?

Continuity Versus Discontinuity

If indeed something does exist always everywhere (including apparently empty space), then the essence of existence is continuous. At the same time, to make sense of the diversity in nature, the continuity of that which exists must vary, from place to place and from time to time. These variations are interpreted as discontinuities in matter and energy and are called particles. But what separates these discontinuities cannot be absolute nothingness, for energy is everywhere always. If the sea is the energy, the sea waves are the fluctuations of the energy, that is, the discrete particles of matter and the discrete particles of force. But even between the sea waves there exists water, the sea, energy, not nothing. So the view of modern physics is some kind of combination of the Parmenidean Being (of an indivisible, continuous whole obeying one eternal truth), the Heraclitean constant change (of everything in the sensible world), and the Democritean discreteness (of a whole, which while in essence continuous is also inhomogeneous and discrete, for it fluctuates). Now what exists must, we believe, be describable by a single idea or equation, a single type of particle, a theory of everything. Can the human intellect ever conceive it? What is the role of the senses in conceiving it?

Intellect Versus Senses

The taste of honey is relative (subjective) for Democritus, and the passage of time is relative (subjective) for Einstein. But both believe reality is objective. By contrast, the creators of the Copenhagen interpretation, Bohr and Heisenberg, think the moon exists only when we look at it; thus, reality is subjective. But Democritus also thought that reality is much deeper than what's revealed by sense perception alone. Trying to capture both the unreliability and the significance of sense perception in our attempts to understand nature rationally, Democritus imagined a hypothetical dialogue between the intellect and the senses.

Intellect: "Sweet is by convention, bitter by convention, hot by convention, cold by convention, color by convention, in reality however there are but atoms and the void."[25]

Senses: "Troubled Intellect! From us you take the evidence and you want to overthrow us? Our fall will be your fall."[26]

---

[25] Sextus Empiricus, *Against the Professors* 7.135–37, trans. Demetris Nicolaides. Or see Graham, *Texts of Early Greek Philosophy*, 595 (text 136).
[26] Galen, *On Medical Experience* 15.7, trans. Demetris Nicolaides. Or see Graham, *Texts of Early Greek Philosophy*, 597 (text 139).

The *Intellect* says that what's perceived by the *Senses* is radically different from the way nature really is. Knowledge derived by the *Senses* is "obscure"[27] but by the *Intellect* "authentic"[28] (Democritus said). The *Senses* perceive sight, hearing, smell, taste, and touch, but these sensations are not objective properties of nature. They are only perceptions by convention (in relation to us, emergent properties), only a consequence of the atoms and their motion in the void—the objective truth, in other words, the true nature of things, is only atoms and the void the *Intellect* claims.

This might be true, the *Senses* respond, but though unreliable ("obscure"), the quest for knowledge always begins with sense perception. For the evidence of the atoms and the void is obtained through observation of colors, tastes, and so on, thus the participation of the *Senses*. It is what we see that we use in order to explore what we cannot see, the *Senses* emphasize. After all, "the phenomena [what is seen, appearances, occurrences] are a sight of the unseen,"[29] said Anaxagoras. At the end, neither the intellect alone nor the senses alone can lead to the truth, but their combination might.

Conclusion

Leucippus's and Democritus's notion of indivisible (atomic), discrete particles without substructure has endured and, according to modern physics, is still one of the most remarkable properties of nature. Could space and time have an atomic nature, too?

---

[27] Sextus Empiricus, *Against the Professors* 7.138–139, trans. Demetris Nicolaides. Or see Graham, *Texts of Early Greek Philosophy*, 597 (text 140).

[28] Ibid.

[29] Ibid., 7.140, trans. Demetris Nicolaides. Or see Graham, *Texts of Early Greek Philosophy*, 309 (text 63).

13

Atoms of Space and Time

Introduction

Epicurus (341–270 BCE) expanded on the atomic theory propounded by Leucippus and Democritus. He wrote three hundred books on just about everything, but unfortunately most have been lost. Only the Stoic philosopher Chrysippus (ca. 279–ca. 206 BCE) wrote as many. According to legend,[1] at fourteen years old he was taught from Hesiod's *Theogony* (on the origin of the gods) that "In the beginning Chaos came into being"[2] . . . which later gave birth to other things. Where did *Chaos* come from?, Epicurus inquired. Baffled by the question, his teacher finally replied, saying that such a query was philosophical. Then I want only philosophers as my teachers, Epicurus responded.

New Features

The Epicurean theory of atoms has several unique features: (1) The E-atom (Epicurean atom), like the D-atom, is unsplittable, but unlike the D-atom, an E-atom has distinguishable indivisible (uncuttable) parts—it has substructure. (2) In addition to material atoms (smallest cuts of matter), there exist space "atoms" (smallest spatial *expanses*) and time "atoms" (smallest time *intervals*)! (3) The motion of an E-atom is dotted, quantum! It moves from one place to the next but *without passing through any of the points in between.* (4) An E-atom may spontaneously swerve (creating uncertainty in its whereabouts), a feature added by Epicurus in a first-ever attempt to escape the Democritean determinism and subject human free will to a scientific hypothesis—the subject of the next chapter.

These news features evolved when Epicurus or his students responded to various well-thought Aristotelian attacks on Democritean atomism. Space and time atoms, we'll see, challenge profoundly the bedrock of modern physics. Space atoms, recall, are an essential requirement of loop quantum gravity. The cause of the most consequential premise of quantum mechanics—the Heisenberg uncertainty principle—and also the cause of the famous quantum jump will

---

[1] Diogenes Laërtius, *The Biography of Epicurus* 10.2–3.
[2] Hesiod, *Theogony* 116.

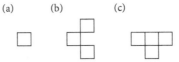

Figure 13.1 (a) a part; (b) E-atom with three parts; (c) E-atom with four parts.

be cautiously speculated with two original ideas, using the Epicurean theory of matter, space, time, and motion.

The Parts of E-Atoms

E-atoms are physically indivisible but *conceptually*[3] divisible into minimal parts. That is, if we could zoom in on an E-atom, we would notice a substructure, that it has physically distinguishable features, identical parts (although their shape and size is not specified in the extant sources). A part can neither be cut further (into smaller pieces) nor can it be cut from (or added to) an E-atom. It is also fundamental (a part is partless): if we continue zooming into a part, we'll discover no new parts (no new substructure). Since we can't really see one, a part is only a mentally identifiable base unit. Different kinds of E-atoms are made from their own unique *integral* number of parts (although not from one part)[4] arranged in multitude ways. Figure 13.1 shows a hypothetical part, and two types of E-atoms in two dimensions. So the range of allowed sizes (and overall geometry) of E-atoms is *quantum*—for example, 2 parts, 3, 4, 5, . . . , but not, say, 4.7 parts—like the range of energies in the modern chemical atom (or matter in general). By contrast, the size of D-atoms is not quantum—it varies continuously.

Democritus speculated the existence of different atomic shapes and sizes with hooks and crevices to explain how D-atoms interact and aggregate to create macroscopic things. Epicurus, by contrast, needed only speculate the notion of a part. Thereafter he imagined parts to have always been connected with each other in various ways, creating thus the varied atomic shapes and sizes.

But atoms couldn't move unless space, time, and motion are radically reimagined.

---

[3] Other words often used are *mentally, theoretically*, and *mathematically*.
[4] Aristotle argued that a partless thing can't move.

Quantum Motion, Space Atoms, and Time Atoms

In criticizing Democritean atomism, Aristotle had argued rigorously that a D-atom, which is partless, can't move unless (a) its motion is *discontinuous, quantum,* and unless (b) space and time are composed of atoms of space and atoms of time—thus, they, too, have a quantum nature.[5]

Concerning notion (a): Aristotle imagined a D-atom to *have moved* from here to there but *without moving through any of the points in between—exactly* the meaning of a *quantum jump* in modern quantum physics (recall the chessboard-pawn analogy)! Incidentally, the Bohr model of the hydrogen atom (recall chapter 6) was the first place to have assumed the exact same type of motion when its electron was making a transition, a quantum jump, from one allowed orbit to the next.

Concerning notion (b): he argued that, if (a) were to hold, then one *should not* imagine that there were ever-smaller regions of space so small as to be points of space, and that there were ever-briefer periods of time so brief as to be moments. Instead, if (a) were to hold, then both space and time should be *finitely divisible*— that is, their division (say, in your mind) can't go on forever; it should stop at a final cut.[6] That is, there should be space and time minima (smallest cuts), *indivisible* spatial and temporal *magnitudes, atoms* (the meaning of *indivisibles* in Greek) *of space* and *atoms of time*! So space *shouldn't* be made of points but from minimal *expanses,* and time *shouldn't* be made of moments (pass moment after moment) but from minimal *periods* (pass period after period).

But Aristotle had also claimed to have successfully shown that all these ideas were impossible. For example, he thought quantum motion (jumps), that something could *have moved* without *moving,* is absurd. Hence, for him there can't be atoms of matter; there can be only a plenum (i.e., matter forms a continuum without a void). Motion can't be quantum; it must be smooth, continuous, sequential point by point (i.e., as we experience it daily). There can't be atoms of space; space is divisible ad infinitum. And there can't be atoms of time; time flows smoothly, moment after moment. Thus, atomism, Aristotle thought, is a mere chimera. But Epicurus is braver. He meets the Aristotelean criticism and boldly embraces exactly *all these "absurdities"*![7] There *are* atoms (of matter, of space, and of time) and motion *is* quantum. Aristotle's reasoning is a force! Even when we

---

[5] Aristotle, *Physics Z.* It is beyond the purpose of this book to present the details of Aristotle's complex arguments. The interested reader may find a great analysis in the book by David J. Furley, *Two Studies in the Greek Atomists* (Princeton, NJ: Princeton University Press, 1967), chap. 8.

[6] Aristotle argued that quantum motion requires that space and time atoms exist. Now, in modern quantum mechanics, motion is quantum; thus, with Aristotle's thinking in mind, a new quantum-mechanical theory should require that space and time atoms exist, too.

[7] Furley, *Two Studies in the Greek Atomists,* 113–116.

think it might be wrong, it's been shaping up history and what we think might be right.

Why Should Atoms Exist?

Epicurus, like Democritus and Leucippus, insisted on matter-atoms and the void because only then motion (and thus change in nature) made sense. Also, the poet Lucretius (99–55 BCE), an ardent Epicurean who preserved many Epicurean ideas through a long poem, lists a variety of observational evidence in support of the atomic hypothesis.[8] But a theoretical argument that guided Democritus to atomism had to do with the nature of divisibility of matter (although not of space and time, but since Democritus's general argument applies to space and time, we'll include that, too). He argued divisibility had to be finite—it had to stop at some minimal nonzero *magnitudes*. If it didn't, if divisibility were infinite, "eventually," he thought, matter (and space and time, everything) will be reduced to nothing—to zero-size points.[9] How could one then reconstruct the universe starting from nothing? Being does not come from Not-Being. Lucretius's Epicurus adds, "If things were made out of nothing, any species would spring from any source and nothing would require seed. Men could arise from the sea and scaly fish from the earth . . . ,"[10] and all sorts of other random things would happen, too. But we have order: "Again, why do we see roses appear in spring, grain in summer's heat, grapes under the spell of autumn?"[11] Order, for Lucretius's Epicurus, was additional evidence that Not-Being cannot generate Being—thus what is has always been and will always be. And for Being to exist and be reconstructible from fundamentals, divisibility had to stop at a final nonzero cut, giving us a material atom (a smallest bit of matter that couldn't be zero), a space atom (a smallest expanse of space that couldn't be zero), and a time atom (a smallest period of time that couldn't be zero). Parenthetically, the strings of string theory and the space atoms of loop quantum gravity have minimal, nonzero magnitudes, too.

Nature, for Epicurus is quantum (finitely divisible) in all its essence: not just as a property of matter (and energy), as our current laws state, but as a property of space and time, as some of our most advanced scientific hypotheses claim. With

[8] Lucretius, *On the Nature of the Universe*, Books One and Two, trans. R. E. Latham (London: Penguin Books, 2005).

[9] Although Anaxagoras thinks otherwise. Infinite divisibility doesn't *ever* lead to an actual *smallest* magnitude (e.g., of zero size) because what is, Being, can never seize to be, cannot become Not-Being. It rather leads always to a *smaller* magnitude (as also implied by Zeno's paradoxes). See Simplicius, *Physics* 164.17.

[10] Lucretius, *On the Nature of the Universe*, 1.159–162 (14).

[11] Ibid., 1.175–176 (14).

space and time atoms, the philosopher had set sail to uncharted waters with unprecedented consequences that shake sacred modern physics grounds.

Space-Time: Points and Smoothness Versus Magnitudes and Quantum-ness

Matter atoms (D-atoms, E-atoms, even the QL-atoms) are separated by a spatial interval, the void—thus, matter spreads discontinuously in space. But nothing separates the space atoms or the time atoms. It wouldn't make sense to say, for example, that space atoms are separated by void for void *itself* is space and with space atoms Epicurus described space itself.

Quantum Space

Space in Newtonian and quantum physics, or even in the more advanced notion of space-time, of relativity, is a smooth, sea-like continuum made of points (thus, it is infinitely divisible). Space in E-physics forms a continuum, too. But it is a granular, a *quantum* space continuum, made not of points but from connected indivisible identical minimal *expanses*, the space atoms (thus, space is finitely divisible). (The size and shape of space atoms is not specified in the surviving sources.) Think of granular space as asphalt-like (with the recognizable imperfections of the asphalt as space atoms) or as a Lego structure (with indivisible equal-size Lego pieces as space atoms), or a Minecraft structure (with its indivisible unit blocks as the space atoms), like Picasso's cubism, a tiled floor, or the game Tetris. Fascinatingly, the space atoms of loop quantum gravity, too, form a granular space and are imagined as interlinked rings.[12] Also, in both E-theory and loop theory the granular space entails a quantum geometry—only certain distinct shapes and sizes are allowed. In a simplified example, if E-space is a tiled floor (figure 13.2) and space atoms are the tiles (of side say 1 unit), the allowed (say square) areas are only $1 \times 1 = 1$ square tile, $2 \times 2 = 4$ tiles, $3 \times 3 = 9$, $4 \times 4 = 16$, and so on. Square areas with numbers in between—such as the "dark gray" area of $2.5 \times 2.5 = 6.25$— are not allowed; "dark gray" areas are allowed only in a point-by-point space. Recall, in relativity, *geometry* is gravity. If geometry is quantum, gravity must be quantum, too. This is a basic hypothesis in loop quantum gravity.

---

[12] Carlo Rovelli, *Seven Brief Lessons on Physics* (New York: RiverHead Books, 2016), 43 (Kindle ed.).

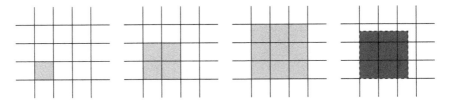

Figure 13.2　Space is quantized in E-theory. Only certain discrete (quantum) areas (geometrical shapes in general) are allowed: the "gray," but not the "dark gray."

Quantum Time

Time in relativity is relative, but it is classical, not quantum. It flows smoothly, moment after moment (as time flows in Newtonian and quantum physics, too), and all moments are allowed to happen. But E-time is quantum! It passes granularly, period after period, time atom after time atom, and so only certain distinct time durations are allowed to happen. Time atoms are indivisible (*not* made from moments), identical, minimal *periods* (although their duration is not stated in the extant writings). Say a time atom is 2 seconds. Time is quantum because only periods 2, 4, 6, 8, . . . seconds occur, all other in-between periods (e.g., 2.5, 7, or 17.2 seconds) are not allowed—they are the "dark gray" periods. But time is quantum also because of the way time passes. In our example, the 2-second time atom passes "whole," not moment by moment—otherwise there would be a problem (see section "Zeno Can Finally Get the Door," later)—for moments can't happen, only durations can: a time atom is indivisible; it is not made from smaller "cuts," or moments. Analogously, the parts of E-atoms aren't made from other parts (or material points), and space atoms aren't made from points of space. That's basically what is meant by the notion of "indivisibility" of a time atom, a space atom, and the E-part. You may think of this as Aristotle thought of a mathematical line, that such a line *contains* points but it is not *composed* of points—for, the nonzero magnitude of a line cannot be reconstructed from points which, by definition, have zero magnitude. As seen earlier, it's difficult to make a universe from points.

Quantum Events

An event in E-physics happens at a spatial expanse, the space atom, and lasts as long as a period of time, a time atom—only when the time atom is up can there be a change into another event. Events therefore are reimagined as space-time *magnitudes* in E-physics, not as space-time points (as in relativity). And as in quantum physics, the E-events are discrete, quantum, not continuous as in relativity. Events in E-physics also can't be "singular." For there are neither moments (of zero time duration) nor points of space (of zero size), not points of matter. Thus, matter can't accumulate at a singularity—an infinitesimally small, a *true* point of space, of size zero—for space is made of *magnitudes*, not of *points*. Nor could the universe have begun as a big bang singularity, but it could as a big bang *magnitude*. Singularities in modern physics are undesirable; they pose existential threats. They are as problematic as suggested by the indeterminate answer of 1/0. Their elimination would be a scientific and mathematical breakthrough if we ever manage it. Only time can tell, if it's not dead.

Is Old Chronos Dead?

On April 6, 1922, during a dialogue between the great philosopher Henri Bergson (1859–1941) and Einstein, the great physicist famously provoked the philosopher by saying that "the time of the philosophers did not exist."[13] He said that in the context of his relativity for which time "flows" deterministically, and the future has already been written (since all moments of time exist always in the block universe). However, the philosopher insisted that "the future is in reality open, unpredictable, and indeterminate."[14] Determinism versus indeterminism, as properties of the universe, is still very much an open question and will be discussed further in the next chapter in the context of free will.

For now, if space and time atoms turn out to be the new reality of nature, it is the time of the physicist that will not exist. For time and space, in Einstein's relativity, although *relative* (i.e., not universal) and revolutionary, are not made of time and space atoms; they are made of points.

Nonetheless, time could be dead, really dead, for both physics and philosophy, but for a different reason. Time, in loop quantum gravity, is hypothesized to not exist—it's an emergent property (and it's not part of the theory's main equations).[15] Physicist Carlo Rovelli (1956–), a founder of loop theory, writes: "For a hypothetically supersensible being, there would be no 'flowing' of time: the universe would be a single block of past, present, and future. But due to the limitations of our consciousness we perceive only a blurred vision of the world and live in time."[16] "Similarly, [in Epicurean philosophy] time by itself does not exist"[17];

---

[13] Einstein quoted in Jimena Canales, *The Physicist and the Philosopher: Einstein, Bergson, and the Debate That Changed Our Understanding of Time* (Princeton, NJ: Princeton University Press, 2016), 5.
[14] Bergson quoted in ibid., 45.
[15] Rovelli, *Seven Brief Lessons on Physics*.
[16] Ibid., 62.
[17] Lucretius, *On the Nature of the Universe* 1.459 (21).

it is an emergent property for "nature is twofold, consisting of . . . matter and the space [void]."[18]

Is Einstein's idea of a cosmic, absolutely constant speed (*not just for light*), required by his relativity, a truth for E-atoms, too?

Cosmic Speed

E-atoms move always with the same constant speed "as quick as thought."[19] It is so because for Epicurus, "Empty space [or in modern physics, the absence of the Higgs field] can offer no resistance to any object in any quarter at any time, so as not to yield free passage as its own nature demands. Therefore, through undisturbed vacuum [before the Higgs field was activated] all bodies [E-atoms, QL-atoms, photons] must travel at equal speed."[20] Epicurus, therefore, according to Lucretius, justifies the constancy of the E-atom speed as a consequence of the void, which offers no resistance to the motion of E-atoms. Analogously, in modern physics, during the early superhot stages of the expanding universe, before the Higgs field was activated, nature allowed QL-atoms to move with their natural cosmic speed, that of light. They slowed down only when the universe cooled a bit and triggered the Higgs field that endowed QL-atoms with their mass (an emergent not a fundamental property) via the Higgs mechanism.

The constancy of speed in E-physics can be justified also in another way. Assuming they move, E-atoms *must* move with the same *constant* speed since all space atoms have the same size and all time atoms last the same interval of time. An E-atom then takes 3 time atoms to travel 3 space atoms, and 5 time atoms to travel 5 space atoms, and so on. Its speed is always constant, 1 space atom per time atom. Speed constancy is inherent in the very quantum nature of Epicurean space and time.

Interestingly, Einstein too thought that *space and time are* the cause that *all* objects move *with a constant speed, that of light*! "Einstein proclaimed that all objects [*not just light*] in the universe are *always* traveling through space-time [*space-time*] at one fixed speed—that of light [even those that appear to move slower]."[21] He means "an object's combined speed through *all four* dimensions— three space and one time— . . . that is [always] equal to that of light."[22] The

[18] Ibid., 1.504-05 (22).
[19] Diogenes Laërtius (Epicurus's), *Letter to Herodotus* 10.61.
[20] Lucretius, *On the Nature of the Universe* 2.234–237 (43).
[21] Brian Greene, *The Elegant Universe: Superstrings, Hidden Dimensions, and the Quest for the Ultimate Theory* (New York: W. W. Norton & Company, 1999), 50.
[22] Ibid. See also World Science U, the courses on relativity for an animated and masterful explanation of this subtle point, http://www.worldscienceu.com/courses/university (accessed July 16, 2019).

consistency of both relativity and Epicurean physics relies subtly on speed constancy! The phenomenon of time dilation, in relativity, implies that time passes ever-slower at ever-greater speeds, and stops at the speed of light. Now, since, in the sense just discussed, everything moves with the speed of light, then time (old Chronos) might indeed be dead.

The constancy of the speed of light (of photons) is a foundational premise in Einstein's special relativity. It safeguards causality and saves the universe from paradoxes (recall chapter 6). Photons travel with the same constant cosmic speed *always*: when they are emitted from moving or stationary sources, when they collide with other particles and change direction, even when they move through other substances. E-atoms too move in the same exact manner (although not with the speed of light).

However, both E-atoms and photons *appear* to move slower when they move within a material substance such as water or oil because of their collisions with other particles. The denser the material, the slower their speed; or more precisely, the slower their speed *appears* to be. This is so because as E-atoms or photons are making their way from one place to the next within a substance they constantly collide with the matter particles of the substance (the other E-atoms, the QL-atoms). Hence, they execute a zigzag, *longer* motion, take longer to arrive at their destination, and as a result appear to have a slower speed. Just like crossing a crowded room with your usual constant pace: zigzagging around the people in an effort to traverse from the back of the room to the front, gives someone who times your trip the impression that you were moving slower. Photons produced in the center of the sun, for example, take a million years to reach its surface because of collisions and zigzagging. Thus, they appear to have been travelling with a speed much smaller than c, although always, from collision to collision, in the sun or any other substance, they are traveling with c. If the sun were a void (empty), the photons would reach its surface in just a few seconds.

Now despite that all E-atoms have the same speed, Epicurus didn't ignore the obvious: that some things are motionless, others move slow, and others fast. But he thought "slowness and quickness are just the appearance which collision and non-collision take on,"[23] that such difference in speed was merely an emergent (an apparent, not a fundamental) property of composite things, and he explained it via the following ingenious mechanism (of particle interaction).[24] Composite objects are made from E-atoms, which all have the same cosmic speed. If it happens that the motion of the E-atoms is random (i.e., those moving in one direction are as many as those moving in the opposite direction), their atomic

[23] Diogenes Laërtius (Epicurus's), *Letter to Herodotus* 10.46, trans. Furley, *Two Studies in the Greek Atomists*, 125.
[24] Furley, *Two Studies in the Greek Atomists*, 123–125.

motion averages out to zero and (while every E-atom in the object is in constant motion) the object itself, as a whole, remains at rest—for example, the E-atoms that push the object, say, right are as many as those that push it left, keeping the bulk of the object at rest. (This is, by the way, the exact same reason that water in a cup doesn't leap out despite that each water molecule in it is in constant motion.) Now, if but only a few of these E-atoms happen to move in a preferred direction, say right, then these E-toms create a small net push to the right causing the object to move slowly to the right, too. The object's motion is increasingly faster in a certain direction the more E-atoms happen to move in that direction. But could E-atoms move in various directions?

Cosmic Direction

At first, the answer appears to be no because E-atoms were imagined to move toward some cosmic, absolute direction, "down"—not down toward the earth. I'm absolutely lost, still, however hard I have labored to find Epicurus's cosmic direction—the extant works don't elaborate much where this direction might be. Nonetheless, I cannot help imagining absolute direction as something like the experience I have whenever I stare at some of the art of M. C. Escher (1898–1972), follow its pathways, feel certain I'm ascending while suddenly I'm also descending. Especially the 1960 print, *Ascending and Descending*, which is "showing a staircase that goes on endlessly ascending—or descending, if you see it that way."[25] Maybe some strange space-time continuum could accommodate an absolute direction, but I don't know!

The Challenge to Make a Universe with Uniform Motion

How could E-atoms ever meet, collide, aggregate, and make a universe of composite things if they always move with a uniform motion, that is, with the same cosmic speed and in the same absolute (cosmic) direction? In the spirit of this book, to read the past from the perspective of modern knowledge, not to really judge it for its weakness, but to be inspired by its insights, I will attempt first a different answer than Epicurus's.

They couldn't meet if space is flat, Newtonian-like. In an analogy, imagine space as a flat chessboard. Flat surfaces obey the principles of Euclidean geometry. Since parallel lines (as those of the chessboard) in flat surfaces never meet,

---

[25] M. C. Escher, *29 Master Prints* (New York: Harry N. Abrams, 1983), 22.

E-atoms with their uniform motion could never meet either. (a) A trailing atom moving along a chessboard line couldn't overtake a front atom for they move in the same direction with the same speed. Also (b) they couldn't meet when two E-atoms move in the same direction along two parallel lines because parallel lines, we learned from Euclid's *Elements*, his superb geometry book, never meet in flat surfaces.

But for case (b) they could meet if the E-atoms move in a curved space, like the surface of, say, a sphere. For according to the principles of Riemannian geometry that describe curved surfaces, parallel lines could meet: two (earth) meridians are parallel as they cross the equator of the earth (or a sphere) but do meet at the north and south poles. Hence, E-atoms moving in the same direction, along different meridians, can collide where the meridians meet, interact, and make a universe.

Space in relativity is curved, but Epicurus didn't know about curved space so he devised another, interesting explanation.

Atomic Swerve

With uniform motion E-atoms can never catch up with each other, collide, interact, and aggregate into composite things. Epicurus addresses the problem by imagining that E-atoms *swerve spontaneously*. He thought that the swerve solved two problems. (1) It allowed E-atoms to change direction, interact, and aggregate into stars, planets, and people. (2) The spontaneity of the swerve introduced uncertainty in atomic motion, a property that Epicurus thought restored human free will, the topic of the next chapter. Swerving has received the intense criticism from both ancient and modern scholars for Epicurus had never explained it; he had merely postulated it—swerve was cause-less for Epicurus. To the best of my knowledge, swerving has not been explained yet. What is the cause of the famous Epicurean swerve? I'll speculate.

First Cause: The Pause

In my opinion the swerve doesn't have to be postulated. Its cause is basically the time atom. First, strike the requirement of the absolute direction of motion (we'll consider it later). Consequently, the quantization of motion, of space, and of time allows E-atoms to move randomly and swerve. How so?

D-atoms move continuously, smoothly point by point, moment by moment— just as we experience the motion of the everyday. Their flow is absolutely uninterrupted, their momentum directional, deterministic.

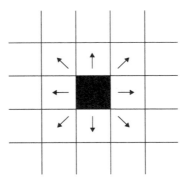

Figure 13.3 The pause of an E-atom is the cause of its spontaneous quantum swerve.

But E-atoms differ. Although their speed is constant—they always cover the same distance in a given time—their motion is *discontinuous*; they *pause*! While an E-atom is in a space atom (at some "location"), it *pauses*, and it is at rest there for a period of a time atom—time, recall, passes period by period not moment by moment. This period of stasis—the time atom—is the cause of motion randomness that Epicurus so much wanted as a property of the E-atom. For, when an E-atom's time atom is up, *and* with the absence of cosmic directionality, its next quantum move (to a neighboring space atom) *must* be random, uncertain: in the absence of directionality, and while on stasis, it has no more reason to move in this neighboring empty space atom ("location") than that one, so it swerves! The swerve is caused by the pause, the time atom, which in essence breaks the flow, the momentum, the directionality and continuity of motion. Time in Epicurean physics is quantum, composed not of "nows" (moments) but of periods, of time atoms entailing a time pause, the cause of the swerve.

In an analogy, a tiled floor (figure 13.3) is a two-dimensional quantum space, and the tiles are space atoms. Imagine a black square E-atom[26] occupying a tile. Since (a) the E-atom is at rest (at a pause) there for exactly a period of a time atom, since (b) it must quantum jump (for its speed is constant, though not continuous), since (c) it lacks directionality (cosmic or from a continuous momentum) it has no more reason to jump here than there—arrows indicate the potential jumps—then (d) when its time atom is up and must jump, it jumps randomly into *any* of the eight tiles around it; thus, it swerves.[27] Its motion is a

[26] Color is simply for reference. Also E-atoms can't be composed from just one part (as Aristotle argued). So the square partless E-atom is an innocent simplification that does not change the essence of this particular discussion.
[27] Detail: if the atom was already moving, say, right, then it has 1/8 chance to continue so; thus, it will not swerve. But the directions in real space are so many that the chance to continue in the same direction is astronomically small and anyway it can't happen always.

constant quantum swerve constituted by uncertain jumps causing uncertainty in its position and direction of motion. Position and direction uncertainties are also part of the famous Heisenberg uncertainty principle. What's the connection?

Cause of Heisenberg Uncertainty:
A Hypothesis Involving No Collision

My hypothesis is that the swerve (and the resulting position-direction uncertainty) is a property inherent in the very fabric of the quantum nature of Epicurean time. Based on that I speculate that time quantization, the time atom, the pause, might be the cause of the Heisenberg uncertainty principle, too—more precisely, all (a), (b), (c), and (d) properties, earlier, contribute. In other words, the position uncertainty of, say, an electron (or generally, a QL-atom), as described by Heisenberg's uncertainty principle, is really caused by the pause—time quantization. Reinforcing my speculation is yet another known truth of the Heisenberg uncertainty principle. That the position-velocity (thus position-direction, for velocity is really speed with direction) uncertainty of an electron is not caused only when we curious observers decide to observe little things by shining photons on them and cause collisions (as discussed in chapter 7). Rather, it is always true, even when we are not observing—an inherent property of nature which now makes more sense in the context of time atoms (time quantization). An electron, I hypothesize, like an E-atom (of figure 13.3), quantum swerves into uncertainty *on its own*—that is, it obeys the Heisenberg uncertainty principle—because of its pause (time quantization).

Second Cause: Overcrowding Is Against the Law

Let's use the tiled-floor analogy to explain the swerve but now by preserving the cosmic direction of motion. For even now E-atoms can meet and interact. For example, with the Riemannian geometry of a sphere in mind, two E-atoms may move in the same direction, north, and still come face to face as long as one moves along the 0-degree meridian, and the other along the 180-degree meridian.

So a black E-atom, which comes from left (figure 13.4.a), moves toward a gray E-atom, which comes from right. The arrows indicate where the atoms have come from. If the E-atoms are initially an odd number of tiles apart, they will eventually face off with a single tile in between them. What then? They pause the usual period of a time atom and when that time is up, each has to quantum swerve to a *different* tile (figure 13.4.b shows one possibility) simply because overcrowding

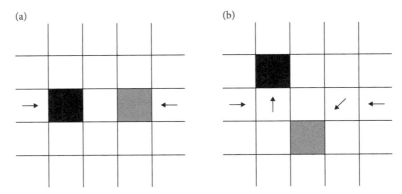

Figure 13.4 (a) After two E-atoms have "collided" (interacted), (b) they must quantum swerve because overcrowding is against the law; they can't occupy the same tile at the same time.

is against the law (Pauli exclusion principle): E-atoms are absolutely rigid; they can't occupy the same tile at the same time.

In a variation, E-atoms will still quantum swerve even if initially they are an even number of tiles apart—only now, they'll swerve after they face off by being right next to each other.

Cause of Heisenberg Uncertainty:
A Hypothesis Involving Collision

The face-off between E-atoms basically describes a collision between them, even if they don't touch. Electrons collide in this manner, for electrons too "know" that overcrowding is not allowed. And as we learned, a collision is the first step of the uncertainty principle. True for E-atoms, too, for after they face off, which tile each will jump into is uncertain. It is as uncertain as an electron's quantum jumps: for, due to the Heisenberg uncertainty principle, an electron (or any microscopic particle) may jump/be only in one of several allowed positions. Here is a more concrete connection: imagine the black E-atom to be an electron and the gray E-atom to be a probing photon used to "see," locate the electron (figure 13.4.a). The Heisenberg uncertainty in the position and direction of the electron is easily justified within the Epicurean physics by figure 13.4.b: after the photon collides with the electron, the electron (the black E-atom) must *quantum jump* into any one of the eight neighboring tiles around it (or to other nonneighboring ones), but which one it is uncertain— thus, both its position and direction are uncertain. Specifically, the distance between the initial tile and the tile that the electron jumps into is the uncertainty in its position.

Particle Claustrophobia

The Heisenberg uncertainty principle is the cause for particle claustrophobia. The smaller the confining space of, say, an electron, the faster it moves away from it—the less time it stays there. The constancy of the speed of E-atoms is the cause of *their* claustrophobia. The E-atom of figure 13.3 stays on one tile (a small confining space) only for the period of a time atom (short period of time), but in a vicinity of say the eight tiles (a bigger confining space) that surround it, it stays at least two such periods (for the E-atom can move back and forth). Could there be a connection between these two seemingly different causes of claustrophobia especially in light of Einstein's realization that in the four-dimensional space-time all objects move with the same speed always? We have no answer so far. However, the realization that claustrophobia—a phenomenon of Heisenberg uncertainty—is also explainable in the context of E-theory (of space and time quantization) is to me another hint that the Heisenberg uncertainty might be caused by space and time quantization.

Challenges

Additional challenges in the Epicurean theory deal with how the size and shape of an E-atom (or its parts) relate to those of a space atom. Can an E-atom, for example, underflow, fit exactly, or overflow a space atom? Epicurus thought an E-atom must fit exactly in the space atoms it occupies. For our examples the E-atom is either *totally* within a tile or not at all: it can't be *partially* in it. Or, can a quantum jump occur into a faraway tile, skipping the immediate next? Relativity imagines objects attached on the fabric of space-time. Since loop quantum gravity attempts to reconcile relativity and quantum mechanics, I anticipate similar challenges between loop quantum gravity's space atoms and QL-atoms.

Zeno Can Finally Get the Door

With space and time atoms, *and* quantum motion, Zeno's paradoxes may be addressed more easily. Imagine there exist five space atoms (five tiles in our simplified quantum space) between here and the door. Zeno (an E-atom) quantum jumps from tile to tile and in a mere five jumps and only five time atoms (five

(a)   (b)

Figure 13.5 (a) An E-atom in continuous motion would always be moving into a space already occupied by itself—an impossibility because E-atoms are impenetrable. (b) But an E-atom in quantum motion (as in the chessboard-pawn analogy) can never move into a space already occupied by itself. Thus a possible cause of an E-atom's (or generally a QL-atom's) quantum jump may be its impenetrability.

periods of time) he finally gets the door, and (dichotomy) problem solved! That's roughly the solution proposed by loop quantum gravity, which assumes space atoms exist.[28]

Note that the paradox would not have been resolved if motion is not quantum, that is, if an E-atom is assumed sliding smoothly, passing sequentially via every point. For such case an E-atom would have Zeno's original challenge (created by space divisibility ad infinitum) to cover any distance such as that of a single space atom (tile). Moreover, if motion were continuous, then part of an E-atom's body (shaped like an arrow in figure 13.5, for simplicity) would have always been moving into the space already occupied by the E-atom itself (figure 13.5.a), but that's impossible because E-atoms are impenetrable; they can't move where they already are because overcrowding is not allowed. Thus, an E-atom *must* quantum jump, as in Figure 13.5b: the arrow has moved to its dotted position without moving through any of the points (locations) in between (since these points were already occupied by itself). Interestingly, this type of reasoning explains the cause, the why QL-atoms, too, must quantum jump; because they, too, are impenetrable.

Time atoms also solve a time-like dichotomy paradox (mine, not Zeno's). First, assume time is infinitely divisible (i.e., no time atoms). To have, say, 1 second pass, time must first flow half a second, then half of the remaining time, then half of the new remaining time, ad infinitum. Since there will always exist a smaller last half to flow/pass last, 1 second can never pass. It's a paradox because I've been growing older by the second.

A second can pass, however, if we suppose time is finitely divisible, that there exist, say, half-second time atoms *and* that time passes time-atom after time-atom—that is, the time interval of the time atom passes *whole, not* moment by moment. One second has passed when first, the first half-second atom

[28] Carlo Rovelli, *Reality Is Not What It Seems* (New York: RiverHead Books, 2017), 217 (Kindle ed.).

has passed, whole, and then the second half-second atom has passed, whole, too. That is, the time duration of 1 second *has passed* (half-second after half-second) *without passing* (from each moment for there aren't moments; there are only periods). So we can't have either a "dark gray" square or a "dark gray" time duration—for our example, ¼, ¾, or 3¼ seconds are "dark gray" durations and can't occur, but 3 or 3½ seconds are allowed.

Epicurus and Planck

How big are the space and time atoms? The Epicureans simply said they are small. Max Planck (1858–1947) speculated an answer (although I'm not sure if he studied Epicurean philosophy). For fun, he played around with the three most important, fundamental constants of nature, the speed of light, c (a cosmic speed limit); the gravitational constant, G (a measure of the strength of gravity); and Planck's constant, h (a measure of the microscopic scale where quantum effects become measurable), by rearranging them in two different ratios. One ratio expresses length, now known as Planck length, an unimaginably small expanse of space, 10^{-35} meters; the other ratio expresses time, known as Planck time, an unimaginably small period of time, 10^{-43} seconds. If space and time turn out to be composed of such small magnitudes, their smallness would make their effects unnoticeable in everyday phenomena. We would be naturally tricked by it and think that space and time are smooth, made of points of space and moments of time—not granular, made of finite magnitudes, the space and time atoms. Planck length is speculated to be the size of space atoms in loop quantum gravity and the size of the strings in string theory.

Conclusion

There are astounding conceptual similarities not only between Epicurean physics and the verified laws of modern physics but also between various Epicurean scientific hypotheses and the hypotheses of cutting-edge physics. I have no doubt that Epicurean and Aristotelian reasoning, especially the arguments regarding space and time atoms, can truly be instrumental in our search for a theory of everything. It is really sad that not too many books have survived from Epicurus but also from Democritus—another prolific writer on just about everything. Is this fate or could it have been avoided?

14
It's Fate—Maybe

Introduction

Since everything is made of atoms (elementary particles in general), then the behavior of all living things must be dictated by the laws that these atoms obey, the laws of nature, Democritus thought. In fact, no biological system hitherto has violated this general realization. Schrödinger's book *What Is Life?* alludes to that by contemplating the physics of living cells and of life. Of course, nowadays the specialized fields of biochemistry and biophysics have been constantly verifying this idea. Does such a realization, however, mean that the laws of nature have an influence on human free will? That is, when I decide to eat pizza instead of Chinese food, have *I* really decided what to eat, or has "my decision" been predetermined by the laws that nature obeys? Generally, do we have free will or not?

Determinism Versus Indeterminism

The physics of Democritus, Newton, Maxwell, and Einstein (his relativity) is deterministic; by contrast, the physics of Epicurus (because of the swerve) and of quantum mechanics (because of Heisenberg uncertainty principle) is probabilistic. Einstein's relativity is the better theory for describing the world of the large, and quantum mechanics is better for describing the world of the tiny. And recall that we lack a unified description of nature.

Determinism means that the motion of every atom in the universe is precise (even if we can't ever know it). How atoms moved in the past determines precisely how they will collide and move in the future—a "rigidity" in "motion." So my lunch future, *if* it can be connected to atomic behavior (and motion rigidity), is, say, pizza because it has been predetermined for me by how atoms moved, collided, and interacted in the distant past (before I existed) and every moment thereafter, until some of these atoms ended up forming me and my brain and making me feel hungry for pizza, which I ended up having. Clarification: determinism doesn't mean that someone who's done a bad deed shouldn't really be held accountable since such a deed has been predetermined by the universe and so it was no one's fault. It rather means that the very action of someone being or not being held accountable *itself* is an *involuntary* action; it's not a choice at all, and it too would have been predetermined by the laws of nature (not of the courts)—because for a deterministic universe there is an unalterable chain of causes, fate.

Indeterminism, on the other hand, means that the motion of every atom in the universe is uncertain and probabilistic. How atoms moved in the past is uncertain and thus how they will move in the future will be uncertain, too. An atom may move this or that way, in general, and it may move in one of several allowed ways—a "fluidity" in "motion." So my lunch future, *again* if it can be connected to atomic behavior (and motion fluidity), will be one from a rich menu of possibilities: pizza, Chinese, or something else. If, say, the atoms moved this way, they could form a brain that wants pizza; if they moved that way, they could form a brain that wants Chinese. But say I ended up eating pizza, was it a choice or chance?

Does Indeterminism Save Free Will?

In other words, does indeterminism imply free will? Epicurus realized that D-atoms are deterministic and reasoned that that meant lack of free will. But unlike the Stoics, he didn't want "to be a slave of the 'fate' "[1] and so he decided to fix it. He was the first to attempt a rational explanation of free will within the context of a scientific theory. He thought his spontaneous swerve (which added fluidity in the motion of E-atoms) restored free will, ethics, and the idea of personal responsibility for our actions. But did it? How free is free will really, if the swerve, which is supposed to be causing it, is really *spontaneous* and thus can't be controlled? "*When the atoms are traveling . . . at quite indeterminate times and places they swerve.*"[2] If we can't control atomic swerving, how can we control our actions, which are supposed to be the result of the swerve? The swerve (or its modern version, the Heisenberg uncertainty principle) *does* give E-atoms extra ways of moving, which may be correlated to a variety of lunch potentialities (pizza, Chinese, something else), but if we can't control the swerve, what good is it in claiming control over "choices"? What I will end up eating might just be the result of chance (of quantum probability or the spontaneity of the swerve)—pure luck—not choice. So we are not sure if indeterminism (of modern or Epicurean physics) entails free will. Would things be different if we could control the swerve

---

[1] Diogenes Laërtius (Epicurus's), *Letter to Menoeceus* 10.134, trans. David J. Furley, *Two Studies in the Greek Atomists* (Princeton, NJ: Princeton University Press, 1967), 174.

[2] Lucretius, *On the Nature of the Universe* 2.218–219, trans. R. E. Latham (London: Penguin Books, 2005), 43.

(or the indeterminacy of quantum mechanics), our luck? In other words, does the well-known folk saying that hard work brings better luck have any scientific basis? Let's see.

Indeterminism in quantum physics means that we can't control the outcome of an experiment. The very act of observation, of interference in general, we learned (in chapter 7), causes uncertainty in atomic motion. But what if we *could* somehow control the swerve (in order to make atoms move a particular way in hopes of controlling our choices). If we could do that, we would be violating the very freedom (the fluidity in motion) that atoms supposedly should have if *we* were to have freedom to control them (and in turn, control our choices). That is, controlling the swerve, which is meant to lead to free will, creates an antinomy, a contradiction: (1) atoms have freedom (in the way they move because they swerve), which (2) *supposedly* gives *us* the freedom to *control them back*, but this (controlling back) (3) also removes their very freedom which supposedly is required—it violates the spontaneity in their swerve that they should have in the first place—if *we* were to *have* freedom.

Thus, it appears that, while indeterminism (via the swerve or the uncertainty principle) may be connected to more lunch potentialities, it doesn't necessarily entail free will—the actuality of what I'll end up eating might just be the result of chance, not choice.

Does Indeterminism Kill Free Will?

But could indeterminism entail *lack* of free will, *determinism*, a rigid regularity? It's a strange, paradoxical question that I'm prompted to ask because of an interesting observation: that macroscopic determinism (order and regularity) arises out of microscopic indeterminism (the uncertainty and probability of quantum mechanics).[3] That is, although the constituent particles of, say, the sun, a flower, our brain, obey randomness, quantum indeterminacy, these *large* (macroscopic) objects (made of *lots* of atoms) obey order and predictability. Does such macroscopic order (regularity) and predictability imply lack of free will?

First, a quick analogy of macroscopic determinism arising from microscopic uncertainty. Say you've been observing Manhattan for a month from a zeppelin high up, using just your eyes. You will discover that lots of people visit the beautiful city daily. Although you can't predict if one particular person will be visiting Manhattan on the next regular day, you can easily predict that lots of people will.[4]

---

[3] Erwin Schrödinger, *What Is Life? & Mind and Matter* (Cambridge: Cambridge University Press, 1967).

[4] For more details of this type of analogy, see Bertrand Russell, *Religion and Science* (New York: Oxford University Press, 1997), 154.

That is, although the laws of quantum mechanics don't allow us to predict an individual atom's exact behavior, the laws of quantum *statistical* mechanics allow us to make reasonable predictions for the behavior of lots of atoms—and macroscopic objects are made from lots of atoms. We can't predict when one radioactive carbon-14 nucleus will decay; but if we have millions of them, we can determine pretty precisely that half of them will decay in 6,000 years. Thus, although each and every atom of say a planet, of the sun, and of our body follows the laws of quantum indeterminacy, the *collective* behavior of all the atoms in a system (the behavior of a macroscopic object, which is made from many particles) is *determinable*: the earth spins once a day, our heart beats seventy beats per minute, tapping a cellphone app opens it, and our energy-hungry complex brain, which although comprising only 2 percent of our body weight, requires 20 percent of the energy from the food we consume.

Here is now the important question: is there a regularity (regardless if it's obvious or not) in the way the brain thinks, as there is in the way it consumes energy (or in the way the earth spins)? If such regularity is discovered to exist, that might imply lack of free will. Because in the same way we know that the earth will be here in the summer and there in the winter in relation to the sun, and the brain will consume so much energy by tomorrow, we would also know that the brain (we) will think this particular way next week, Monday, noon (even with each brain atom's actions being uncertain). Of course, note, if such regularity exists but can't be figured out, our next Monday's action, although predetermined, would still be unknown to us and thus appear to be our choice. Now, while there *is* a certain regularity (and predictability) in the way someone thinks, such regularity is not absolute. For example, I know that I'll have my usual coffee tomorrow morning, and also I have always known that I would study physics. But I may not have my usual coffee tomorrow, and it might not have worked out to study physics. This uncertainty (whatever its source might be) *is* of course the critical debate of the issue and what gives free will another chance. How?

Part Free Will and Part No Choice

According to quantum statistical mechanics, the regularity of a system in general becomes increasingly more precise with increasingly more particles present in the system. So it's easier to predict the collective action of the many (e.g., of the general vote of millions of people, or when half from a billion carbon-14 nuclei will decay) than the action of the one. Now, since the universe is the ultimate many-particle system, as a whole (or on a large scale), it appears to be rather deterministic (regular) and predictable, too, in its behavior—that is, we know it is expanding, how stars form and die, and how colliding black holes ripple space-time. On the other hand, it is impossible to predict the behavior of the individual microscopic particles (electrons, protons, etc.) that constitute the universe. It appears, then, that determinism might be as much of a property of the universe as its opposite, indeterminism. That might be true for us, too—how we think and act. This is because we are much smaller than the universe, so we are not perfectly deterministic, but at the same time, we are much bigger than an electron, so we are not perfectly indeterministic. Determinacy might govern us as much as indeterminacy—we might be part free will and part no choice. That would be fair, cosmically just, as Anaximander might have put it, I'm certain, perhaps. Besides, as argued in Empedocles's cosmological cycles (chapter 10), for a universe with no beginning or end, with "first" and "last" to lack absolute meaning, the properties of the composite (of the many and large) are as fundamental as the properties of the few and tiny.

Conclusion

In summary, what's on trial here is whether macroscopic order and predictability imply complete or partial lack of free will, or more generally, whether determinism or indeterminism found in our scientific theories implies anything at all concerning free will. The verdict: I think that's unanswerable, but that's a blessing, not a blemish. I personally feel, though unscientifically (without evidence), that I have free will, yet again, it might have been predetermined that I feel so. If Fate does exist, what would be determining her decisions, and the decisions of that? Indeterminism (via the swerve or its modern version, the uncertainty principle) implies free volition no more than determinism does. The debate on free will, as old as Aristotle's treatises on ethics, has not been settled yet. Not only because we lack a theory of everything (which may be deterministic, probabilistic, or some combination of the two concepts) but also because for either interpretation of nature, arguments can be made for and against voluntary action. Einstein was disenchanted by the quantum indeterminacy as implied by his famous "God doesn't play dice," an indirect attack on quantum mechanics' founding principle, the Heisenberg uncertainty. With this principle there was a lot at stake for Einstein because the philosophy of his relativity is deterministic, the exact opposite of the quantum indeterminacy. Regardless of how history is being written (deterministically, via chance, via free will, or some combination) it is time that "we" consider it.

15

Atomic Connections

Epicureanism has arguably been an intellectual bridge between ancient and modern science. It contributed to the spreading of Leucippus's and Democritus's atomism but also of Epicurus's own unique rendition of it, to later thinkers.[1]

For example, the great Latin poet Lucretius (99–55 BCE), a devoted Epicurean, composed the epic didactic poem *On the Nature of Things*,[2] which expounded masterfully the Epicurean philosophy that, according to Roman orator Cicero (106–43 BCE), has "taken over the whole of Italy."[3] Epicureanism "survived, it is true, though with diminishing vigour, for six hundred years after the death of Epicurus."[4] It "was a significant trend in Hellenistic times [300 BCE–200 CE]"[5] and among those who studied it (and atomism in general) during this period were the brilliant scholars of the Library of Alexandria.

In addition, "the echoes of this battle [between the atomists and their critics (e.g., the Platonists, Aristotelians, and Stoics) who held that matter is continuous] were heard from time to time in medieval [Middle Ages (500–1500)] Europe, and it flared up again with great intensity in the sixteenth and seventeenth centuries."[6] Then, Epicureanism was revived and analyzed with even greater zeal when a copy of Lucretius's poem resurfaced in 1417 in a monastery, inspiring anew various Renaissance philosophers. These included priest Giordano Bruno (1548–1600), who sadly was burned alive at the stake because his scientific ideas were considered heretical by the Roman Inquisition. Like Copernicus and Galileo, his cosmology was heliocentric, not geocentric, as Aristotle and the Catholic Church held. Astronomer, physicist, and self-proclaimed philosopher Galileo cited Lucretius's work to compare the Epicurean physics of falling bodies with Aristotle's[7] and his own.[8] (Legend has it that he first tested his hypothesis of

---

[1] Stephen Greenblatt, *The Swerve: How the World Became Modern* (New York: W. W. Norton, 2011).

[2] Also known as *On the Nature of the Universe.*

[3] Mentioned in the modern introductory commentary of Lucretius, *On the Nature of the Universe*, trans. R. E. Latham (London: Penguin Books, 2005), xxiii.

[4] Bertrand Russell, *The History of Western Philosophy* (New York: Simon & Schuster, 1945), 251.

[5] Andrew Gregory, *Eureka! The Birth of Science* (Cambridge: Icon Books, 2001), 138.

[6] David J. Furley, *Two Studies in the Greek Atomists* (Princeton, NJ: Princeton University Press, 1967), v.

[7] Lucretius, *On the Nature of the Universe* 2.226–227 (43), 2.156 (41n15); compare Epicurus's and Aristotle's theories of falling bodies.

[8] Daniel Kolac and Garrett Thomson, *The Longman Standard History of Philosophy* (New York: Pearson, 2005), 253.

falling bodies by dropping objects from the Leaning Tower of Pisa, but later by actually building an inclined plane in order to slow down motion and measure it more easily, a now standard freshman physics experiment.) He included this work in the *Discourses and Mathematical Demonstrations Relating to Two New Sciences*, a physics book he published in 1638 at the age of 74 that transformed the physics and mathematics of motion. This book inspired Newton, an atomist himself, who in turn inspired Einstein, who had proven atomism theoretically, and who inspired everyone after him. Philosopher priest Pierre Gassendi (1592–1655) "definitely reintroduced atomism into modern science [by promoting Epicureanism against the then-preferred Aristotelianism], and he came to it after studying the fairly substantial extant writings of Epicurus."[9] "From the lives and writings of Gassendi and [philosopher, mathematician,[10] scientist René] Descartes [(1596–1650)], who introduced atomism into modern science, we know as an actual historical fact that, in doing so, they were fully aware of taking up the history of the ancient philosophers whose scripts they had diligently studied."[11] Philosopher, physicist, and chemist Robert Boyle (1627–1691) studied Epicurean physics, too—incidentally, Boyle's gas law is also a standard freshman physics experiment.

Ancient atomism continued to be learned by Enlightenment philosophers such as the empiricist John Locke (1632–1704); polymath Roger Joseph Boscovich (1711–1787), who published a treatise about atomism and forces in 1758; and Founding Father of the United States Thomas Jefferson (1743–1826), who proclaimed, "I too am an Epicurean."[12] Since "there are some interesting similarities between the Epicurean theory of indivisibles [atom, recall, means indivisible] and Hume's theory of space and time [part II of Hume's *A Treatise of Human Nature*, one of Einstein's favorite books],"[13] and moreover since David Hume (1711–1776) attributed the so-called Epicurean paradox (on the why there is evil in the world) to Epicurus, I'm inclined to suppose that another influential thinker, Hume, was a student of Epicureanism.

One of the challenges for accepting atomism was the debate on the void, which was required to separate the atoms and allow them to move. Plato and Aristotle, for example, whose intellectual influence has practically been uninterrupted through the ages since their beginnings, rejected atomism and embraced the plenum. But as regards science, in the end it matters not who you are and

---

[9] Erwin Schrödinger, *Nature and the Greeks and Science and Humanism* (Cambridge: Cambridge University Press, 1996), 75.

[10] The Cartesian geometry (of the x-y coordinate system) studied by all math students is named so because Cartesius is Latin for Descartes.

[11] Schrödinger, *Nature and the Greeks*, 82–83.

[12] Anthony Grafton, Glenn W. Most, and Salvatore Settis (eds.), *The Classical Tradition* (Cambridge, MA: The Belknap Press of Harvard University Press, 2010), 323.

[13] Furley, *Two Studies in the Greek Atomists*, 136.

what you teach because the single most important judge for the truth of a scientific hypothesis is experiment, observation, *evidence*! Science is knowledge based on *evidence*.

The first evidence for the atomic nature of matter came relatively recently, in the nineteenth century, with the experiments of John Dalton (1766–1844). (Atomic evidence awaited 2,500 years for the "hand" [technology] to catch up with the ideas of the mind and build the proper tools for experimentation.) It was then that atomism began its transition from a purely theoretical scientific hypothesis into an experimentally verified one—thus into a law of nature! The influential physicists Maxwell and Ludwig Boltzmann (1844–1906) embraced atomism, too. Incidentally, Maxwell's theoretical work on electromagnetic undulations (caused by accelerating discrete, atomic particles bearing electric charge) inspired Einstein to imagine (in his relativity) space-time undulations— also known as gravitational waves, detected recently as a result of colliding black holes—caused by matter. The atomic concept was established further with additional experimental work from other scientists, including physicist J. J. Thomson, who discovered the electron in 1897.

In 1905 Einstein predicted theoretically the existence of the atom when he explained Brownian motion[14] by requiring that his equations treat matter as atomic. Interestingly, Leucippus and Democritus required that matter is atomic in order to explain rationally motion, too, any motion. The year 1905 was Einstein's miracle year: he also published his theory of special relativity, the equivalence between mass and energy (the famous $E = mc^2$), and the explanation of the photoelectric effect by treating both matter (the electrons) *and* energy (light, the photons), as atomic (made of discrete particles, discontinuously distributed in space). His explanation of the photoelectric effect won him the Nobel Prize in Physics in 1921.

Thomson's student physicist Ernest Rutherford discovered experimentally the atomic nucleus in 1911. Rutherford's student Niels Bohr advanced further the theoretical notions of an atom (with his famous Bohr model of the hydrogen atom) and so did the research of so many others, experimental and theoretical scientists, including Heisenberg, Schrödinger, Paul Dirac (1902–1984), Feynman, and *every other physicist* from the twentieth century on, for atomism has since been an undisputed law of nature and has been incorporated (directly or indirectly) in our teachings and our research.

In fact, the physicists of loop quantum gravity aspire to demonstrate that atomic are not just matter and energy, but also space. I'm certain their research

---

[14] The zigzag motion of a tiny particle submerged in a fluid.

will benefit from the rigorous logic of Epicurus and of his students—in fact, even from the logic of Aristotle (via his well-argued criticism of atomism).

The influence of an idea doesn't come only from its supporters, but from its intense critics, too. There haven't been more fierce critics of atomism than the Eleatic school of thought (of Parmenides and Zeno), the Platonic, the Aristotelian, and the Stoic. And who, from the great thinkers of the Renaissance, the Enlightenment, and later did not read those philosophical schools, especially the last three? Even Friedrich Nietzsche (1844–1900) who, although "admired the pre-Socratics [the Greek natural philosophers from the sixth and fifth centuries BCE] except Pythagoras,"[15] rejected atomism, and so did the renowned philosopher and physicist Mach, who declared, "I do not believe that atoms exist,"[16] and chemist Wilhelm Ostwald (1853–1932) were also spreading it, for their philosophies have been so well scrutinized.

"If I have seen further it is by standing on the shoulders of Giants," stated Newton in 1675. Newton studied Epicurean physics,[17] Latin,[18] Greek, and Euclidean geometry. It's not just natural philosophy or in general science that spread ancient atomism from the past to the present; it is also mathematics simply because from the time of Pythagoras it had been realized that the language of science is mathematics. For example, "Plato's main contribution to science sprang from his realization of the problem of the irrational [numbers and lines], and from the modification of Pythagoreanism and atomism."[19] Inspired by the Pythagorean mathematics, Plato evolved into a *geometrical* atomist[20] (as seen in chapter 6) by replacing the Pythagorean "things are numbers" with things are shapes, forms, *Forms*.

In light of this super brief history of ideas, it is clear that the intellectual continuity between ancient natural philosophy and modern physics has undoubtedly been tight and systematic.[21] I'm hoping with this book that this relationship will

[15]   Russell, *History of Western Philosophy*, 776.
[16]   Carlo Rovelli, *Reality Is Not What It Seems* (New York: RiverHead Books, 2017), 41 (Kindle ed.).
[17]   Daniel Kolac and Garrett Thomson, *The Longman Standard History of Philosophy* (New York: Pearson, 2005), 253.
[18]   His famous *Philosophiae Naturalis Principia Mathematica* (Mathematical Principles of Natural Philosophy) was written in Latin.
[19]   Karl R. Popper, *Conjectures and Refutations: The Growth of Scientific Knowledge* (London: Routledge, 1989), 87.
[20]   Ibid., 81; Gregory, *Eureka*, 55–59.
[21]   Bruce Thornton, *Greek Ways: How the Greeks Created Western Civilization* (San Francisco: Encounter Books, 2002); Carl Sagan, *Cosmos* (New York: Random House, 1980), chap. 7; G. E. R. Lloyd, *Early Greek Science: Thales to Aristotle* (New York: W. W. Norton & Company, 1970); Gregory, *Eureka*; Russell, *History of Western Philosophy*; Isaac Asimov, *The Greeks; A Great Adventure* (Boston: Houghton Mifflin, 1965); Leon Lederman and Dick Teresi, *The God Particle: If the Universe Is the Answer, What Is the Question?* (Boston: Houghton Mifflin, 1993), chap. 2; Popper, *Conjectures and Refutations*; Rovelli, *Reality Is Not What It Seems*; Schrödinger, *Nature and the Greeks*, 3–99; Stephen Bertman, *The Genesis of Science: The Story of Greek Imagination* (New York: Prometheus Books, 2010); Werner Heisenberg, *Physics and Philosophy: The Revolution in Modern Science* (New York: Harper Torchbooks, 1962).

be revived not only for history's sake but also for the inspiration of new ideas. I believe that an interplay between philosophy (ancient or modern) and modern physics is always a more illuminating path to the truth. Great philosophers have been physicists and great physicists have been philosophers. In chapter 1, we asked what is philosophy and what is physics? Philosophy is part physics and physics is part philosophy, and like the parts of the E-atom, they (should) constitute an unsplittable union in the search for a theory of everything. A *vision* about nature, a *philosophy*, should be driving the mathematics of physics, not lifeless mathematics forcing itself onto eventful universe.

Epilogue

Our ancient quest for knowledge began with our evolution as a species 200,000 years ago, and with everything we experienced through our struggles to survive and our efforts to thrive and live fully. We hunted and gathered, painted on caves, told stories, domesticated animals and plants, built homes, wondered about nature, gave birth to civilization and religion, picked up writing, philosophized, and engaged in science. In fact, we keep on doing all these wonderful things, but, amazingly, we have been doing them ever more in the light of science.

Since the birth of Greek natural philosophy 2,600 years ago, science has evolved significantly. Nonetheless, its ultimate goal still remains essentially the same: to understand nature rationally and to reduce the explanations of all natural phenomena to the least possible number of basic assumptions (first causes, axioms)—ideally to just one, hence a unified theory of everything. Now, say that has been achieved, will the human intellect be satisfied?

We like Homer's *Odyssey* so much because it is a story of a journey (in fact, a long one), not of a destination. Shortly after Odysseus returns to Ithaca, the story ends and we feel melancholic—we like the journey better than the destination. Although, with Odysseus's return, the events in Ithaca were breathtakingly exciting and awaited eagerly from the start of the story, their completion also brought the absolute end of the epic adventure.

The beauty of nature is in her secrets; the magic is in our discoveries. I never want to know everything—to have the journey of knowledge, of search and discovery, ever end. What would be next if I did? What would happen to the magic? Space, time, matter, energy, the human senses to observe, and the intellect to contemplate—it is all nature, and her nature is her many secrets (her shadows). They are many but also intelligible (steal-able, like Promethean fire)! I hope you have a magical, endless journey searching, in the light of science, for the nature of nature!

A student of Euclid once asked, What would I earn if I learned geometry? If you should earn from what you learn, here is a coin, Euclid responded sardonically. But, really, why should we learn? Because we can, might be the best answer, but also because the journey of knowledge is infinitely interesting. Nonetheless, in front of the wisdom of the universe, we should remain always humble, for as many learned people have come to know (Democritus, Socrates, Russell), we might be wrong on just about everything.

Bibliography

Asimov, Isaac. *Asimov's Chronology of Science and Discovery*. New York: HarperCollins, 1989.

Asimov, Isaac. *Asimov's Chronology of the World*. New York: HarperCollins, 1991.

Asimov, Isaac. *The Greeks; A Great Adventure*. Boston: Houghton Mifflin, 1965.

Asimov, Isaac. *Understanding Physics*. US: Dorset Press, 1988.

Baker, Joanne. *50 Physics Ideas You Really Need to Know*. London: Quercus, 2007.

Bertman, Stephen. *The Eight Pillars of Greek Wisdom*. New York: Barnes & Noble, 2007.

Bertman, Stephen. *The Genesis of Science: The Story of Greek Imagination*. Kindle ed. Amherst, NY: Prometheus Books, 2010.

Boardman, John, Jasper Griffin, and Oswyn Murray, eds. *The Oxford Illustrated History of Greece and the Hellenistic World*. Oxford: Oxford University Press, 1986.

Brunschwig, Jacques, and Geoffrey E. R. Lloyd. *Greek Thought: A Guide to Classical Knowledge*. Cambridge, MA: Belknap Press of Harvard University Press, 2000.

Brunschwig, Jacques, and Geoffrey E. R. Lloyd. *A Guide to Greek Thought: Major Figures and Trends*. Cambridge, MA: Belknap Press of Harvard University Press, 2003.

Burckhardt, Jacob. *The Greeks and Greek Civilization*. Edited by Oswyn Murray. Translated by Sheila Stern. New York: St. Martin's Griffin, 1999.

Burkert, Walter. *Greek Religion: Archaic and Classical*. Translated by John Raffan. Malden, MA: Blackwell, 1985.

Burnet, John. *Early Greek Philosophy*. London: A & C Black, 1920.

Canales, Jimena. *The Physicist and the Philosopher: Einstein, Bergson, and the Debate That Changed Our Understanding of Time*. Princeton, NJ: Princeton University Press, 2016.

Cox, Brian, and Jeff Forshaw. *Why Does E=mc2?: (And Why Should We Care?)*. Kindle ed. Boston: Da Capo Press, 2009.

Dalling, Robert. *The Story of Us Humans, From Atoms to Today's Civilization*. New York: iUniverse, 2006.

Davies, P. C. W., and Julian Brown. *Superstrings: A Theory of Everything?* Cambridge: Cambridge University Press, 1992.

Economou, Eleftherios N. *A Short Journey from Quarks to the Universe*. Berlin: Springer, 2011.

Einstein, Albert. *Relativity: The Special and the General Theory*. Kindle ed. Amazon Kindle Direct Publishing, 2011.

Feynman, Richard P. *The Meaning of It All*. New York: Basic Books, 1998.

Feynman, Richard P. *Six Easy Pieces*. New York: Perseus, 1963.

Feynman, Richard P. *Six Not So Easy Pieces*. New York: Perseus, 1963.

Freeman, Charles. *The Greek Achievement: The Foundation of the Western World*. New York: Penguin Books, 2000.

Freeman, Kathleen. *Ancilla to the Pre-Socratic Philosophers*. Cambridge, MA: Harvard University Press, 1996.

Furley, David J. *Two Studies in the Greek Atomists*. Princeton, NJ: Princeton University Press, 1967.

Graham, Daniel W. *Explaining the Cosmos: The Ionian Tradition of Scientific Philosophy.* Princeton, NJ: Princeton University Press, 2006.

Graham, Daniel W. *The Texts of Early Greek Philosophy: The Complete Fragments and Selected Testimonies of the Major Presocratics.* Translated by W. Daniel Graham. Cambridge: Cambridge University Press, 2010.

Graves, Robert. *The Greek Myths.* London: Penguin Group, 1955.

Greene, Brian. *The Elegant Universe: Superstrings, Hidden Dimensions, and the Quest for the Ultimate Theory.* New York: W. W. Norton & Company, 1999.

Greene, Brian. *The Fabric of the Cosmos: Space, Time, and the Texture of Reality.* New York: Vintage, 2005.

Greene, Brian. *The Hidden Reality: Parallel Universes and the Deep Laws of the Cosmos.* New York: Vintage, 2011.

Gregory, Andrew. *Ancient Greek Cosmogony.* Kindle ed. London: Bloomsbury, 2008.

Gregory, Andrew. *Eureka! The Birth of Science.* Cambridge: Icon Books, 2001.

Hamilton, Edith. *The Greek Way.* New York: W. W. Norton & Company, 1930.

Hawking, Stephen. *A Brief History of Time: From the Big Bang to Black Holes.* New York: Bantam Books, 1988.

Heisenberg, Werner. *Physics and Philosophy: The Revolution in Modern Science.* New York: Harper Torchbooks, 1962.

James, Renée C. *Seven Wonders of the Universe: That You Probably Took for Granted.* Baltimore: Johns Hopkins University Press, 2011.

Kaku, Michio. *Einstein's Cosmos: How Albert Einstein's Vision Transformed Our Understanding of Space and Time.* New York: W. W. Norton & Company, 2004.

Kaku, Michio, and Jennifer Trainer Thompson. *Beyond Einstein: The Cosmic Quest for the Theory of the Universe.* Revised ed. New York: Anchor, 1995.

Kirk, G. S., J. E. Raven, and M. Schofield. *The Presocratic Philosophers.* Cambridge: Cambridge University Press, 1983.

Kolac, Daniel, and Garrett Thomson. *The Longman Standard History of Philosophy.* New York: Pearson, 2005.

Lederman, Leon, and Dick Teresi. *The God Particle: If the Universe Is the Answer, What Is the Question?* Boston: Houghton Mifflin, 1993.

Lightman, Alan. *Great Ideas in Physics.* 3rd. ed. New York: McGraw-Hill, 2000.

Lindberg, David C. *The Beginnings of Western Science: The European Scientific Tradition in Philosophical, Religious, and Institutional Context, Prehistory to A.D. 1450.* 2nd ed. Chicago: University of Chicago Press, 2008.

Lindley, David. *Uncertainty: Einstein, Heisenberg, Bohr, and the Struggle for the Soul of Science.* New York: Anchor, 2008.

Lloyd, G. E. R. *Early Greek Science: Thales to Aristotle.* New York: W. W. Norton & Company, 1970.

Lloyd, G. E. R. *Greek Science after Aristotle.* New York: W. W. Norton & Company, 1973.

Lucretius. *On the Nature of the Universe.* Translated by R. E. Latham. London: Penguin Books, 2005.

McKirahan, Richard D. *Philosophy Before Socrates.* Kindle ed. Indianapolis: Hackett, 2010.

Mourelatos, Alexander P. D., ed. *The Pre-Socratics: A Collection of Critical Essays.* Garden City, NY: Doubleday and Company, 1974.

Oppenheimer, Robert J. *Science and the Common Understanding.* New York: Simon & Schuster, 1954.

Pais, Abraham. *Subtle Is the Lord: The Science and the Life of Albert Einstein.* New York: Oxford University Press, 2005.

Pomeroy, Sarah B., Stanley M. Burstein, Walter Donlan, and Jennifer Tolbert Roberts. *A Brief History of Ancient Greece: Politics, Society, and Culture.* 2nd ed. Oxford: Oxford University Press, 2008.

Popper, Karl R. *Conjectures and Refutations: The Growth of Scientific Knowledge.* London: Routledge, 1989.

Randall, Lisa. *Knocking on Heaven's Door: How Physics and Scientific Thinking Illuminate the Universe and the Modern World.* New York: Harper Perennial, 2011.

Randall, Lisa. *Warped Passages: Unraveling the Mysteries of the Universe's Hidden Dimensions.* New York: Ecco, 2005.

Ridley, B. K. *Time, Space and Things.* 2nd ed. Cambridge: Cambridge University Press, 1984.

Rosenblum, Bruce, and Fred Kuttner. *Quantum Enigma: Physics Encounters Consciousness.* 2nd ed. Oxford: Oxford University Press, 2011.

Rovelli, Carlo. *The First Scientist: Anaximander and His Legacy.* Kindle ed. Translated by Marion Lignana Rosenberg. Yardley: Westholme, 2011.

Rovelli, Carlo. *Reality Is Not What It Seems.* Kindle ed. New York: RiverHead Books, 2017.

Rovelli, Carlo. *Seven Brief Lessons on Physics.* Kindle ed. New York: RiverHead Books, 2016.

Russell, Bertrand. *The History of Western Philosophy.* New York: Simon & Schuster, 1945.

Russell, Bertrand. *The Scientific Outlook.* London: Routledge, 2009.

Sagan, Carl. *Cosmos.* New York: Random House, 1980.

Schrödinger, Erwin. *Nature and the Greeks and Science and Humanism.* Cambridge: Cambridge University Press, 1996.

Schrödinger, Erwin. *What Is Life? & Mind and Matter.* Cambridge: Cambridge University Press, 1967.

Sean, Carroll. *The Big Picture: On the Origins of Life, Meaning, and the Universe Itself.* Kindle ed. New York: Penguin, 2016.

Sean, Carroll. *The Particle at the End of the Universe.* New York: Dutton, 2012.

Strogatz, Steven. *The Joy of x: A Guided Tour of Math, from One to Infinity.* Kindle ed. New York: Houghton Mifflin Harcourt, 2012.

Thornton, Bruce. *Greek Ways: How the Greeks Created Western Civilization.* San Francisco: Encounter Books, 2002.

Vlastos, Gregory. *Studies in Greek Philosophy: Volume 1 The Presocratics.* Princeton, NJ: Princeton University Press, 1993.

Waterfield, Robin. *The First Philosophers: The Presocratics and Sophists.* Oxford: Oxford University Press, 2000.

Index

For the benefit of digital users, indexed terms that span two pages (e.g., 52–53) may, on occasion, appear on only one of those pages.

absolute zero, 23–24, 66
Achilles (paradox), 87, 90–91, 93
adaptation, 27–28, 108–9
air, 42–43, 109
    as Empedocles's element, 6, 99–100, 102–3
    in Empedocles's experiment, 106–7
    as primary substance (in Anaximenes philosophy), 27, 30–31, 32, 34
amphibia, 28
Anaxagoras, 6, 16, 75, 98, 100, 109, 110–19passim, 120–21, 134, 138n.9
    and the Copenhagen interpretation, 111–13
    and fractals, 114–16
    and the many-worlds interpretation, 113–14
Anaximander, 5, 17, 18–29passim, 33, 55, 108–9, 155–56
    and apeiron (as primary substance), 18
    on cosmology, 25–26
    and energy and the apeiron, 19
    on evolution, 27
    and the Higgs Particle, 22–23
    on neutrality, 23
    and why matter is more than antimatter, 24–25
Anaximenes, 5–6, 27, 29, 30–34passim, 99–100, 106–7
    and the atomic theory of matter, 32–34
anti-earth, 45
antielectrons, 19. See also positron (s)
antigravity, 128–29. See also repulsive gravity
antileptons, 21
antimatter. See also antiparticle (s)
    and Anaximander's philosophy, 18, 19, 20, 21, 22–25
    and energy, 77–78
    and Heraclitus's philosophy, 57
    in the observable universe, 10, 18, 24–25
antiparticle (s), of antimatter, 19, 20, 21, 22–23, 128. See also antimatter
antiquarks, 21
Antisthenes the Cynic, 88
apeiron (Anaximander's primary substance), 17, 18, 21, 25–26, 27, 29, 109

and energy, 19, 22
and Higgs particle, 22–23
aphelion, and Kepler's harmonic law, 39
Apollo (Greek god), 2, 14
Aristarchus, and the heliocentric model, 25–26, 46–47
Aristotle, 1, 5, 8–9, 25–26, 36–37, 38–39, 42–43, 53, 87, 90, 92–93, 96, 99–100, 121–22, 140, 146–47n.26, 156, 157–60
    on atomic theory, 137–38
    on the cause of motion, 121n.2
arrow paradox, 41, 61–62, 91–92, 93, 94, 95, 96
arrow of time, 103–4
asceticism, 36–37
Asia (Minor), 27
asymmetry between matter and antimatter, 25
atomic theory, 35, 109, 118–19, 120, 135
    and ancient atoms, 120–21
    and force, 100
    and mass, 132
    and Parmenides's Being and Not-Being, 76, 77
    of Plato, 43
    and rarefaction and condensation, 32, 33
    and Richard Feynman, 125
atom (s), 1, 10, 12–13, 30
    ancient, 120–23
    and Bohr, 39
    chemical, 123
    D-atoms (ancient atoms) and QL-atoms (quarks and leptons): similarities and differences, 123–26
    and gravity, 129–30
    of Leucippus and Democritus, 120
    and motion, 27
    and Parmenidean Being, 77
    and rarefaction and condensation, 32
    of space, 12–13, 98, 135–51passim (see also quantum space)
    of time, 12–13, 98, 135–51passim (see also quantum time)
    and void, 33, 77, 120
axiom (s), 31, 40, 86, 89–90, 121, 129, 162

babies (human), and Anaximander's evolution of the species, 28
Babylonians, and eclipses, 16
Bacon, Francis, 3
Being (s) (Parmenidean), 71, 72–86passim, 99–100, 104–5, 124–25, 126, 133, 138
    and the Copenhagen interpretation, 112–13
    and Einstein's block universe, 73–75
    and the many-worlds interpretation, 113–14
Bell's inequality, 82
Bergson, Henri, 141
big bang, 10–11, 64–65, 66, 83, 84, 86, 99, 104–5, 107, 108, 116, 140–41
    and Empedocles's cosmology, 103–4
big crunch, 103–4
black hole (s), 13–14, 79, 155–56, 159
block universe (Einstein's), 73–75, 93, 141
Bohr, Niels
    on the atom, 39, 137, 159
    on reality, 133
Boltzmann, Ludwig, 159
Boscovich, Roger Joseph, 158
Boyle, Robert, 157–58
Bruno, Giordano, 157–58
Burnet, John, on Anaximander's worlds, 26

calculus, 33, 63
    and Zeno's dichotomy paradox, 89–90
carbon-14, 154–56
causality, 8, 85–86
    in classical and quantum theory (physics), 95
    and special relativity, 48, 81–82, 143
cave art, 106
Central Fire, in Pythagorean cosmology, 5, 45–46
centrifuge, 130
chance
    for Empedocles, 102
    and necessity, 102
    in quantum theory, 80–81, 153–54 (see also probability)
change, in Heraclitus's philosophy, 54–71passim
Chaos (or chaos), 114–16, 135
chemistry, and Empedocles's philosophy, 99, 101–2
Chios (Greek island), 15–16
Chronos, 141–43
Chrysippus, 135
cinematography
    and apparent motion, 93–95
    in observing, 62
classical physics, 58–59, 60–61, 80–81, 82, 93, 95–96. See also Newton, Isaac (or Newtonian)

claustrophobia, 149
clepsydra, 6
    in Empedocles's experiment, 106–7
color charge, 23
comet (s), 16
condensation (in Anaximenes philosophy), 30–35passim
consciousness, 86, 141–42
conservation, laws, 20, 24–25, 26, 57, 69–70, 100–1, 123–24
    of electric charge, 20
    of energy, 19–20
    of momentum, 80
Copenhagen (interpretation), 80–81, 82, 97, 133. See also many-worlds (interpretation)
    and Anaxagoras's philosophy, 110, 111–14
Copernicus, Nicolaus, 25–26, 46–48, 51–52, 157–58
cosmic calendar, 104–6
Cosmic Dark Ages, 105, 108
cosmic direction, 144, 146, 147
cosmic inflation, 104–5
cosmic microwave background, 23–24, 105, 106
cosmic speed, 48, 142–44, 151
cosmology. See also big bang
    for Anaximander, 25–26
    cycles in modern, 103–4
    for Empedocles, 102–3, 108
    fractal (or hierarchical), 114
    Pythagorean, 36, 44–45
cosmos, for Pythagoras, 38–39
Croesus, King, 14
Croton, southern Italy, 36–37

Dalton, John, 159
Dante (Alighieri), 15
dark energy, 10, 128–29
dark matter, 10, 128–29. See also matter, ordinary
deductive reasoning, 40
Delphi, 1, 14
Democritus, 5, 12–13, 16, 27, 32, 33, 34, 35, 75, 76, 98, 100, 109, 116–17, 118–19, 120–34passim, 135, 136, 138, 151, 152, 157, 159, 162
    on atoms and the void, 33
Descartes, René, 72–73, 157–58
determinism (classical), 11–12, 60–61, 74, 135, 141, 152–53, 154, 155–56. See also chance; indeterminism; probability
deuterium, 65
dialectic, 1, 6
Dialogue Concerning the Two Chief World Systems (Galilei), 47–48

dichotomy (paradox), 87–90, 149–50
dimensions, in string theory, 12
Dionysus (Greek god), 2
Dirac, Paul, 159
diversity, in Einstein's block universe, 75

earth
    in Anaximander's cosmology, 25–26
    as a primary substance, 42–43, 99
eclipse (s)
    moon (or lunar), 14, 44–45
    solar, 14–15, 60–61
    and Thales, 14–15
Ecphantus, 51–52
Einstein, Albert, 10. See also relativity
    and the block universe, 73–75
    and Mach's principle, 52
    and quantum entanglement, 80
    and relative motion, 51–52
    and theory of everything, 11–12
Elea, 72, 87
electric charge, 9–10
electric force, 9–10, 100, 105, 131
electromagnetic (force), 9–10, 20, 56, 100–1,
    159. See also forces
electromagnetism, 9–10, 57, 76
electrons (as leptons), 7
electroweak (force), 10–11, 53
elements (of Empedocles [earth, water, air,
    fire]), 99
Elements (Euclid's geometry book), 144–45
emergentism, 102–3. See also reductionism
Empedocles, 6, 75, 98, 99–109passim, 110, 116–
    17, 120–21, 131–32, 155–56
    and natural selection, 108–9
    and Olbers's paradox, 107–8
    and the standard model, 100–2
empty space, 19, 32, 34, 76, 77, 108, 118–19,
    120, 122, 126, 127–28, 129, 130–31, 133,
    142, 146. See also vacuum; void
energy. See Anaximander; antimatter;
    conservation; dark energy; equivalence of
    mass and energy; fire; vacuum energy
entropy, 2n.8, 67, 103–4. See also
    thermodynamics, second law of
Epicurus, 5, 12–13, 53, 90, 92–93, 98, 135–
    51passim, 152, 153–54, 157–58, 159–60
Episteme (knowledge), 1, 6
EPR, a thought experiment, 80
equivalence of mass and energy, 50
ether (a purer kind of air), 38, 42–43
Europe, 157–58
event horizon, 13–14
evolution (biological)

    in Anaximander's philosophy, 27
    in Empedocles's philosophy, 108–9
exclusion principle (Pauli), 42, 127–28, 147–48

Fate, in Stoic philosophy, 153–54
Feynman, Richard, 3, 27, 125, 159
field (s), in science, 22–23
fire, in Heraclitus's philosophy, 69–70
forces, the four fundamental forces of
    nature, 7, 11
Forms (Plato's theory of), 42–44
fractal (s), 114–16
free will, 5, 61, 83, 135, 141, 145, 152–56
Furley, David J, 92–93
fusion, 105

Galilei, Galileo, 47–48, 51–52, 80–81, 118,
    121, 157–58
gamma rays, 65
Gassendi, Pierre, 157–58
general relativity. See relativity
geocentric (model), 36, 44–45, 46–48, 51,
    118, 157–58
geodesic, 131
geometric series, 89
geometrical-arithmetical duality, 42. See also
    wave-particle duality
geometry (or geometrical), 12
    of the Epicurean atom, 136
    Euclidean, 144–45, 160, 162
    for Plato, 44
    as property of the universe, 41–42, 43, 86, 104–5
    quantum, 139
    Riemannian, 145, 147
    and space, 63–64, 129, 131
George Washington Bridge, 93–94
Glashow, Sheldon, 10
gluons (the particles of strong force), 56,
    101–2, 124
god (s), and science, 7, 85–86
Grand Unified Theory, 10–11
gravitational constant, 71, 151
graviton (s) (hypothetical particles of gravity),
    56, 124, 125–26, 131
gravity (or gravitation), 7, 10, 56, 57, 71, 76,
    100–1, 103–4, 105–6, 125, 129, 130–31,
    151. See also forces; loop quantum gravity;
    quantum gravity
    in Democritean philosophy, 129–30
    in general relativity, 11–12, 13, 42, 49, 50, 63–64,
    79–80, 104–5, 128–29, 131, 139, 159
    in Newtonian physics, 25–26, 39, 50, 52, 55–56,
    81–82, 100, 130, 131
Greene, Brian, 39–40, 127–28

Halys River, 14
harmonic law (of Kepler), 39
Hawking, Stephen, 2, 3
Heisenberg, Werner, 27, 32n.5, 133, 159
    on Heraclitean fire, 70
    and the uncertainty principle, 54, 57, 58, 60,
        66, 91–93, 127–28, 135–36, 146–47, 148–
        49, 152, 153–54, 156
    on the universal substance, 21–22
heliocentric (model), 25–26, 36, 44–45, 46–48,
    51, 118, 157–58
Heraclides Ponticus, 46–47
Heraclitus, 5–6, 20, 31, 54–71passim, 94–95, 96
Herodotus, 16
Hesiod, 135
Hicetas, 51–52
hierarchy (in cosmology, fractal universe), 114
Higgs, boson, 22, 125–26, 129
    and Anaximander's apeiron, 18, 22–23
    and mass, 10, 22, 128, 132
    and neutrality, 23–24
    and the standard model, 10, 22,
        23–24, 125–26
    and universal substance, 23, 125
Higgs mechanism, 132, 142. See also mass: in
    quantum physics
Higgs, Peter
    and the standard model, 22–23
Hippasus of Metapontum, and the square root
    of two, 41
Homer (or Homeric), 31, 54–55, 162
hominid (s), 28, 117
Homo sapiens, 106, 117
Hubble Space Telescope, 128–29
Hubble's law, 64–65, 106
Hubble, Edwin, 64–65
Hume, David, 158
hylozoism, 14, 27

Iliad (Homer), 54–55
indeterminism (quantum), 60, 61, 141, 152–56.
    See also chance; determinism; probability
inertia, the law of, 61–62, 121, 132
information paradox, 14
Inquisition, 118, 157–58
integers, 37, 39, 40–41
intellect and the senses (a hypothetical
    dialogue), 133
interconnectivity (in the universe), 83
Internet, 106
irrational, number (s), 40, 41, 42–43, 89–90,
    160. See also rational, number (s)
Isis (Egyptian goddess), 16

isotropy (or isotropic, as property of the
    universe), 23–24, 47–48, 65, 121
Italy, 36–37, 157
Ithaca, 162

Jefferson, Thomas, 158
Jupiter, 45

Kepler, Johannes, 39
kinematics, 129

Large Hadron Collider, 10
Lederman, Leon
    on Anaximander's apeiron, 22–23
    on Democritus, 125
    on Empedocles's elements, 100–1
Leibniz, Gottfried, 33
Lemaître, Georges, 10–11
length contraction (in special relativity), 49, 50
leptons (particles of matter), 7, 8, 10, 12, 21–22,
    23, 34, 36, 54, 56, 100–3, 120, 122–23. See
    also atom (s): D-atoms (ancient atoms) and
    QL-atoms (quarks and leptons)
Leucippus (atomic philosopher), 5, 12–13, 32,
    33, 34, 76, 109, 118–19, 120, 121–22, 123–
    24, 126, 129, 134, 135, 138, 157, 159
Library of Alexandria, 157
light, speed of, 19, 48, 49, 50, 52, 56–57, 64, 71,
    81–82, 83–84, 132, 142–43, 151. See also
    relativity
Locke, John, 158
Logos, in Heraclitean philosophy, 57
loop quantum gravity, 12–13, 57, 135–36, 138,
    139, 141–42, 149–50, 151, 159–60. See also
    gravity; quantum gravity
Lorentz transformation, 49
Lucretius, 138, 142, 157–58
Lucy (a species of genus Australopithecus), 117
Lydia (or Lydians), 14–15

Mach, Ernst, 51–52, 160
Mach's principle, 52–53
magnet, 9–10
magnetic (force), 9–10
mammals, 28
many-worlds (interpretation), 110, 113–14. See
    also Copenhagen (interpretation)
mass
    in Democritean philosophy, 129–30, 132
    and Higgs boson (or field or mechanism), 10,
        22–23, 128, 132, 142
    in Newtonian physics, 48, 55, 61–62,
        130, 132

in quantum physics, 13, 21, 39, 56, 58–59, 60, 68, 78–79, 91–92, 94, 123–24, 132 (*see also* Higgs mechanism)

in relativity (special or general), 19–20, 50, 63–64, 77–78, 105, 128, 131, 159

mathematics, limitations of, 31, 40, 86, 89–90

matter, nature of, 30–31, 34, 98, 118–19, 120, 159. *See also* primary substance; universal substance

matter, ordinary, 129. *See also* dark matter

Maxwell, James Clerk, 9–10, 57, 152, 159

Medes (a people), 14–15

Mediterranean, Sea, 16

Mercury (the planet), 45, 46–47

metaphysics, 82

Miletus, 15–16, 27

Milky Way, 25

mind. *See* nous

momentum, 26, 57, 58–59, 61, 80, 81–82, 145, 146–47

monism (or monistic), 27, 71, 72, 76, 109, 116–17

monotheism, 11, 31

moon, 14–15, 38–39, 44–46, 51–52, 60–61, 64, 80–82, 108, 118, 121–22, 133

motion
  absolute, 47–48, 49, 50, 52
  quantum (jump), 92–93, 145–49
  relative, 47–48, 49, 50, 51–52
  unverifiable, 87–98passim

music, and Pythagoras, 36–40, 101–2

music of the spheres, 38–40

mutation, 117

natural selection, 108–9, 118

neutrino, 23, 65

neutron (s), 7, 10, 11–12, 56, 65, 101–2, 123

Newton, Isaac (or Newtonian), 10, 11–12, 25–26, 33, 39, 42, 47–48, 50, 51–52, 54, 55–56, 57, 58–59, 63–64, 73–74, 79, 80–82, 97, 100, 121, 130–31, 139, 140, 144–45, 152, 157–58, 160. *See also* classical physics

Nietzsche, Friedrich, 160

Nile, River, 16

Noether, Emmy, 26

nonexistence, 77–79, 84–85. *See also* nothingness; Parmenides

nothingness, 2, 72–73, 75, 76, 77–79, 100, 126, 133. *See also* nonexistence

nous (mind), 6, 109, 110, 116–18

nuclear strong (force), 7, 10–11, 23, 56, 76, 100–1. *See also* forces

nuclear weak (force), 7, 10, 56, 76, 100–1. *See also* forces

Odysseus, 162

Odyssey (Homer), 162

Olbers's paradox, 107–8

On Nature (Parmenides's poem), 84–85

Oppenheimer, J. Robert, 92–93

Osiris (Egyptian god), 16

Ostwald, Wilhelm, 160

parable of the cave, 4

Parmenides (or Parmenidean), 2, 5–6, 66–67, 71, 72–86passim, 97, 99–100, 102, 109, 112–14, 120–21, 124–25, 126, 127, 128, 129, 133, 160

Parthenon, 117–18

particles, virtual, 128, 129

Pericles, 118

perihelion, and Kepler's harmonic law, 39

periodic table of chemistry, 42, 123

phase transition (s), critical, 114–16

Philolaus, 45

photoelectric effect, 159

photon (s) (the particles of light), 34, 56–57, 58–59, 61–62, 83, 95, 101–2, 122, 123–24, 125, 142, 143, 147, 148–49, 159

pi, 41

Planck, Max, 151

Planck constant, 60, 71, 78–79, 151

Planck length, 12–13, 151

Planck time, 151

Planet of the Apes, 49

Plato, 2, 4, 5, 36–37, 42–44, 52–53, 103–4, 123, 157–59, 160

Platonic Forms, 36, 43–44

Pleiades, 14

Podolsky, Boris, 80

polytheism, 31

Popper, Karl, 2, 3, 25–26, 69–70, 74

positron (s), 19–20, 25, 65. *See also* antielectrons

pre-Socratics, 160

primary substance (s) (of matter, of the universe), 7, 8–9, 11, 12, 17, 18, 21, 23–24, 27, 29, 30, 31, 69–70, 98, 106–7, 109, 110, 125. *See also* matter, nature of; universal substance

primates, 28, 106, 117

probability, 11–12, 42, 60–61, 83–84, 95–96, 102, 111–12, 124–25, 153–54. *See also* chance; determinism; indeterminism

proton (s), 7, 10–12, 56, 58, 61–62, 65, 101–2, 105, 123, 155–56

pyramid (s), 14–15, 106, 117–18

Pythagoras (or Pythagorean [s]), 1, 5, 35, 36–53passim, 83, 101–2, 106–7, 160

Pythagorean theorem, 40

quantum entanglement, 52, 80–84, 130–31
quantum event (s), 140–41
quantum geometry, 139
quantum gravity, 11–13, 139. *See also* gravity; loop quantum gravity
quantum jump. *See* motion
quantum mechanics (or physics, or theory), 11–14, 22–23, 34, 38, 39, 42, 54, 56–57, 58, 61, 63, 66, 68, 77–78, 79, 80–82, 87, 91, 92–93, 95–96, 102, 110, 111–12, 113–14, 124–25, 127–28, 130, 135–36, 137, 137n.6, 139, 140–41, 149, 152, 153–55, 156
quantum numbers (or set), 37, 38, 39, 42
quantum space, 139, 146–47, 149–50. *See also* atom (s): of space
quantum time, 140. *See also* atom (s): of time
quark (s) (particles of matter), 7, 8, 10, 11–12, 21–22, 23, 34, 36, 54, 56, 100–3, 120, 122–23. *See also* atom (s): D-atoms (ancient atoms) and QL-atoms (quarks and leptons)

radioactive decay, 10
rarefaction (in Anaximenes philosophy), 30–35passim
rational, number (s), 40–41, 89–90. *See also* irrational, number (s)
redshift, 64–65
reductionism, 102–3, 122–23. *See also* emergentism
relativity, 2, 3, 12–13, 51–52, 79–80, 139, 140–41, 142–43, 152, 156
    general (and topics related to), 10–12, 13, 14, 36, 42, 50, 51, 57, 63, 64–65, 66, 68–69, 87, 97, 104–5, 128, 130, 131, 139, 145, 149, 159
    special (and topics related to), 8, 19, 47–50, 51, 73–74, 77–78, 81–82, 83–84, 97, 123–24, 143, 159
religion, 11, 31, 36–37, 117–18, 162
Renaissance, 127, 157–58, 160
repulsive gravity, 104–5. *See also* antigravity
Roman Catholic Church, 118
Rosen, Nathan, 80
Rovelli, Carlo, 141–42
Russell, Bertrand, 2, 126–27, 162
Rutherford, Ernest, 123, 159

Sagan, Carl, 33, 105–6
sage (s), 5, 14
Salam, Abdus, 10
Saturn (the planet), 45
Schrödinger equation, 61
Schrödinger's cat, 80–81, 111–13

Schrödinger, Erwin, 27, 32n.5, 34, 80–81, 159
scientific method, 2–3, 36, 41–42, 85–86n.18, 86
Sherrington, Charles, 27
singularity, 84, 86, 116–17, 140–41
Socrates, 1, 5, 162
solar system, 39, 105–6
solidification, 7
solstices, 14
Sommerfeld, Arnold, 39
space atoms. *See* atom (s): of space
space paradox, 96–97
special relativity. *See* relativity
spin (of elementary particles or primary substance), and neutrality, 23–24
spooky action at a distance, 81–82
standard model, of physics, 7, 10–12, 23–24, 56, 57, 99, 100–2, 125–26, 131, 132
Stoic (s), 135, 153–54, 157–58, 160
strife (the force), 99–100, 101–2, 131–32
string, theory (or strings [of string theory]), 12–13, 39–40, 57, 124, 125–26, 129, 138, 151
supernova, 105–6
survival of the fittest, 27–28, 108–9
swerve (atomic, in Epicurean philosophy), 135, 145–49passim, 152, 153–54, 156
symmetry (or symmetric or symmetrical), 26, 31, 43

technology, 117, 118, 159
Thales, 5, 6, 7–17passim, 25–26, 27, 99–100, 106, 125
Theogony (Hesiod's), 135
theory of Forms (Plato's), 42–44
thermodynamics
    second law of, 63, 67, 103–4 (*see also* entropy)
    third law of, 66
Thomson, J. J., 220, 123, 159
time atoms. *See* atom (s): of time
time, relative, 49. *See also* relativity
time dilation, 49, 50, 142–43
time paradox, 97
time-like dichotomy paradox, 150
time travel, 11–12, 49

uncertainty principle (s), 54, 57, 58–60, 61, 63, 66, 68, 77–79, 87, 91–93, 94, 127–28, 149
    and cause of (a hypothesis), 135–36, 146–47, 148–49
    and free will, 152, 153–54, 156
universal gravitation, 10, 52, 130. *See also* gravity

universal substance, 21–22, 24–25, 27, 125.
    See also matter, nature of; primary
    substance
universality, in fractals, 114–16
universe
  age of, 86
  center of, 47–48, 65, 86
  closed, 103–4
  observable (or the "afterglow" of big bang),
    10, 20, 24–25, 105
  open, 103–4
  parallel, 113–14
  size of, 25, 104–5
uranium, 10

vacuum, 96, 127, 128, 142. *See also* empty
    space; void
vacuum energy, 128
Venus (planet), 44–45, 46–47
void, 5, 32–33, 34, 52–53, 76, 77, 97, 98, 106–7,
    120–23, 124, 125–29, 130, 133, 134, 137–38,

139, 141–42, 143, 158–59. *See also* empty
    space; vacuum
vortex, 129–30, 132

W's (W$^+$ and W$^-$, particles of the weak force),
    56, 132
water, as primary substance (in Thales's
    philosophy), 7–9
wave function (s), 36, 42, 43–44, 61, 83,
    95–96, 102
wave-particle duality, 42, 83. *See also*
    geometrical-arithmetical duality
weight
    in Democritus's philosophy, 122, 129–30
    and force in modern physics, 130–32
Weinberg, Steven, 10
Wheeler, John Archibald, 131

Z's (Z^0, particles of the weak force), 56, 132
Zeno, 6, 41, 86, 87–98passim, 138n.9, 140,
    149–51, 160